Revision Chemistry

M.A. Cowd, BSc, PhD, CChem, MRIC
Lecturer in Chemistry,
North Wirral College of Technology

and

P.J. Miller, BSc, CertEd
Head of Chemistry Department,
Ballakermeen School,
Douglas, Isle of Man

(*Both authors were sometime Head of Chemistry Department, Birkenhead Institute High School*)

BUTTERWORTHS
LONDON - BOSTON
Sydney - Wellington - Durban - Toronto

United Kingdom London	**Butterworth & Co (Publishers) Ltd** 88 Kingsway, WC2B 6AB
Australia Sydney	**Butterworths Pty Ltd** 586 Pacific Highway, Chatswood, NSW 2067 Also at Melbourne, Brisbane, Adelaide and Perth
Canada Toronto	**Butterworth & Co (Canada) Ltd** 2265 Midland Avenue, Scarborough, Ontario, M1P 4S1
New Zealand Wellington	**Butterworths of New Zealand Ltd** T & W Young Building, 77–85 Customhouse Quay, 1, CPO Box 472
South Africa Durban	**Butterworth & Co (South Africa) (Pty) Ltd** 152–154 Gale Street
USA Boston	**Butterworth (Publishers) Inc** 10 Tower Office Park, Woburn, Massachusetts 01801

First published 1980

ISBN 0 408 10607 7

British Library Cataloguing in Publication Data

Cowd, M A
Revision chemistry.
1. Chemistry
I. Title II. Miller, P J
540 QD33 79-41164

ISBN 0–408–10607–7

Typeset by Scribe Design, Gillingham, Kent
Printed in England by Camelot Press Ltd, Southampton, Hants.

PREFACE

This book is designed primarily for students studying chemistry to Advanced Level for the examinations set by the Joint Matriculation Board, the Southern Universities Joint Board, London University and the Scottish Certificate of Education Board (Higher Grade). It will therefore cover most of the content of the other major syllabuses – Nuffield, the Associated Examining Board, Oxford and Cambridge, etc. – and also ONC/OND courses, current technicians' courses, and first year chemistry subsidiary courses at colleges and universities. Although there are text books now covering the modernized 'A' level syllabuses, there is a need for a revision note book which can help students with their examination preparations. Furthermore, some schools and colleges may well find this book useful as a concise textbook in itself.

The introductory section presents some fundamental principles in a way which can be easily read. This Introduction has been made deliberately long since some of the more subtle arguments of the syllabuses are presented in this section. Chapters 1–10 cover the physical section of the syllabuses. Chapters 11–13 (inorganic) and Chapters 14–17 (organic) make continual reference to this physical section, thus enabling the authors to keep the content of the inorganic and organic sections to a minimum. These last two sections are treated in a systematic manner, hence avoiding the problem of presenting the student with a large volume of unrelated facts. The inorganic section makes use of the underlying physical principles of atomic structure to explain reactions and also the group and periodic trends; the organic section is treated mechanistically wherever practicable.

Nomenclature is in accordance with the joint statement made by the G.C.E. boards, which is largely that recommended by the Association for Science Education. Physical data used in the text has been taken from the Nuffield Book of Data, and we are grateful to the Nuffield Foundation and the Longman Group Ltd., for permission to do this. We are also grateful to the same bodies for permission to reproduce a diagram on atomic volume and to use some theoretical and experimental values of lattice energy (taken from Nuffield Advanced Science (Chemistry) Teachers' Guide 1). We wish to thank Professor A.K. Holliday and Dr H. Block (both of the University of Liverpool) for their advice and encouragement, and Mr R.B. Jones (formerly of Unilever Research) for reading the physical section of the manuscript.

M.A.C.
P.J.M.

CONTENTS

Introduction 1

1. Nature of Atom and Radioactivity 14
2. Shapes of Molecules 20
3. Forces Between Atoms, Molecules and Ions 24
4. States of Matter 28
   - Gases 28
   - Liquids 34
   - Solids 34
5. Energetics 41
6. Phase Equilibria 49
7. Chemical Equilibria 60
8. Ionic Equilibria 64
9. Electrochemistry 76
10. Kinetics 81
11. Periodicity 87
12. Some Elements and Groups of the Periodic Table 92
    - Hydrogen 92
    - Groups I and II 94
    - Group III 98
    - Group IV 101
    - Group V 108
    - Group VI 115
    - Group VII 120
    - Group 0 125
13. The First Row Transition Elements 127
14. Introduction to Organic Chemistry 142
15. A Study of Some Functional Groups 152
16. Some Biologically Important Organic Compounds 187
17. Synthetic Macromolecules 191

Index 195

INTRODUCTION

There are 103 well defined elements, and it has been shown that these elements can be put into different groups or series, the elements in each group or series possessing similar properties. This is the basis of the Periodic Table. To understand why some elements have similar properties, we must look at the structure of the atoms.

Atomic structure

The atom is found to contain three fundamental particles, the electron, the proton and the neutron; some details of these are given in *Table 1.*

Table 1 Fundamental particles for chemistry

Particle	*Mass relative to hydrogen atom*[a]	*Relative charge*[b]
Proton	1	+1
Neutron	1	0
Electron	1/1840 (often taken as negligible mass)	−1

[a]Rest mass of hydrogen atom is 1.67×10^{-27} kg
[b]Actual charge of proton is 1.602×10^{-19} C

The protons and neutrons are to be found in a dense region of positive charge known as the nucleus, and hence the mass of the atom is concentrated in the nucleus. It can be written:

MASS NUMBER = NO. OF PROTONS + NO. OF NEUTRONS (*see* also Chapter 1)

Since the atom is electrically neutral, then:

NO. OF PROTONS = NO. OF ELECTRONS = ATOMIC NUMBER

The method of showing these properties is to write two numbers to the left of the symbol, the upper one showing the mass number and the lower the proton number or atomic number, thus:

$^{12}_{6}C$ and $^{235}_{92}U$

The electrons are to be found in regions surrounding the nucleus, and information concerning their location can be obtained by determining IONIZATION ENERGIES and examining ATOMIC SPECTRA (*see* later).

If a gas (at low pressure) in a thyratron, a type of radio valve, is bombarded with electrons (usually generated by a heater within the valve), the energy needed to eject an electron from a gaseous atom (known as the ionization energy) can be determined. It is by considering the ionization energies that a picture is built up of electrons occupying main ENERGY LEVELS or SHELLS outside the nucleus. Each shell is capable of holding $2n^2$ electrons, where n is an integer 1, 2, 3 ..... describing the number of the shell. Thus the first shell can hold two electrons, the second eight, the third 18, and so on. As n increases, so the energy of the shells increases – electrons occupying these higher energy shells are, on average, further from the nucleus and are therefore attracted less by the positive nuclear charge.

Within each shell there exist ORBITALS, each orbital holding a maximum of two electrons (a consequence of the Pauli exclusion principle). The first shell contains a single *s* orbital, denoted by the symbol 1*s*. The second shell, however, holds four orbitals, namely an *s* orbital and three *p* orbitals, denoted by the symbols 2*s* and $2p_x$, $2p_y$, and $2p_z$ (note that the first number in all cases refers to the shell number; the subscripts x, y, z indicate that those particular orbitals are pointing along the x, y and z axes respectively). The third shell contains nine orbitals, an *s* orbital, three *p* orbitals and five *d* orbitals, denoted by the symbols 3*s*, $3p_x$, $3p_y$, $3p_z$ and the five 3*d* orbitals. In higher shells there are *f* and *g* orbitals; these will not be dealt with here.

The shapes of these orbitals describe regions of space in which the electrons have a maximum probability of being found (*see* also Chapter 1); the shapes of the *s* and *p* orbitals are shown in *Figure 1.* It should be noted that within any

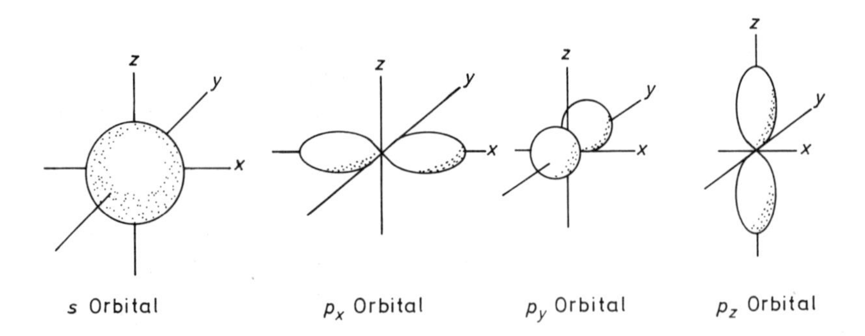

Figure 1 Shapes of s *and* p *orbitals*

one shell, the *p* orbitals stretch further from the nucleus than the *s* orbital and, similarly, the *d* orbitals stretch further from the nucleus than the *p* orbitals; consequently, the order of energy of these orbitals is $d > p > s$ within a shell. Electrons fill up a particular shell, therefore, by first entering orbitals of low energy (for example, an *s* orbital), then eventually entering higher energy orbitals when the low energy ones are full (known as the Aufbau principle). ELECTRONIC CONFIGURATIONS can now be written for the elements, showing how the electrons occupy the orbitals. *Table 2* gives electronic configurations for the elements with atomic numbers 1–10.

Note that in the case of boron onwards, the *p* orbitals fill up singly because of mutual electron repulsion. Only when oxygen is reached do electrons begin to 'pair-up' in the *p* orbitals (in this pairing-up process, the electrons adopt opposite spins). This latter rule (where orbitals of the same energy are occupied singly before pairing occurs) is called Hund's rule.

Table 2 Electronic configurations of the first ten elements

Element	*Electronic configuration*	*Alternative representation** 1s	2s	2p
H	$1s^1$	[↑]		
He	$1s^2$	[↑↓]		
Li	$1s^2 2s^1$	[↑↓]	[↑]	
Be	$1s^2 2s^2$	[↑↓]	[↑↓]	
B	$1s^2 2s^2 2p^1$	[↑↓]	[↑↓]	[↑][ ][ ]
C	$1s^2 2s^2 2p^2$	[↑↓]	[↑↓]	[↑][↑][ ]
N	$1s^2 2s^2 2p^3$	[↑↓]	[↑↓]	[↑][↑][↑]
O	$1s^2 2s^2 2p^4$	[↑↓]	[↑↓]	[↑↓][↑][↑]
F	$1s^2 2s^2 2p^5$	[↑↓]	[↑↓]	[↑↓][↑↓][↑]
Ne	$1s^2 2s^2 2p^6$	[↑↓]	[↑↓]	[↑↓][↑↓][↑↓]

*The boxes represent orbitals

CORRELATION OF IONIZATION ENERGY WITH ELECTRON STRUCTURE

The energy needed to eject the outermost electron from a gaseous atom to infinity is defined as the FIRST IONIZATION ENERGY, I_1, the ejection of the second outermost electron to infinity (the first having been removed) being the SECOND IONIZATION ENERGY, I_2, and so on*. I_2 exceeds I_1, since I_2 involves ejection of an electron from a positively charged ION. The difference in magnitude between I_2 and I_1, however, can depend on several factors; these are now discussed with reference to the eleven successive ionization energies for sodium, atomic number 11, electronic configuration $1s^2 2s^2 2p^6 3s^1$ (*Table 3*).

Column 3 gives the differences in successive ionization energies, and these differences can be explained as follows:

$I_2 - I_1$ and $I_{10} - I_9$ are large because the second and tenth electrons, respectively, are being removed from complete shells.
$I_5 - I_4$ shows the stability of a half-filled set of *p* orbitals.
$I_8 - I_7$ shows the stability of the completely filled *s* orbital.

Information concerning orbitals can also be obtained by plotting I_1 *versus* atomic number (*Figure 2*). *Figure 2* shows that the I_1 values for elements having atomic numbers 2, 10, 18 and 36 (namely He, Ne, Ar and Kr) are high; electronic configurations for these elements are 2, 2.8, 2.8.8 and 2.8.18.8 respectively. It would appear that elements having 8 electrons in their outer shells have particularly stable electronic structures. This eight-electron or 'octet' rule holds for all

*Values are quoted in kJ mol^{-1}

Table 3 Ionization energies of sodium

Ionization energy number	*Ionization energy/* kJ mol^{-1}	*Difference in successive ionization energies/* kJ mol^{-1}	*Electronic configuration after electron removed*
I_1	500		$1s^2 2s^2 2p^6$
I_2	4600	$I_2 - I_1 = 4100$	$1s^2 2s^2 2p^5$
I_3	6900	$I_3 - I_2 = 2300$	$1s^2 2s^2 2p^4$
I_4	9500	$I_4 - I_3 = 2600$	$1s^2 2s^2 2p^3$
I_5	13 400	$I_5 - I_4 = 3900$	$1s^2 2s^2 2p^2$
I_6	16 600	$I_6 - I_5 = 3200$	$1s^2 2s^2 2p^1$
I_7	20 100	$I_7 - I_6 = 3500$	$1s^2 2s^2$
I_8	25 500	$I_8 - I_7 = 5400$	$1s^2 2s^1$
I_9	28 900	$I_9 - I_8 = 3400$	$1s^2$
I_{10}	141 000	$I_{10} - I_9 = 112\,100$	$1s^1$
I_{11}	158 700	$I_{11} - I_{10} = 17\,700$	—

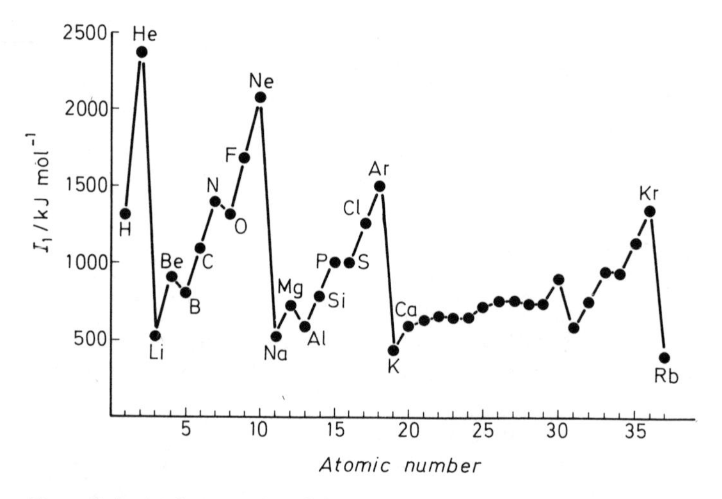

Figure 2 Ionization energies of elements

the noble gases (except He, which has a full 1*s* orbital), and often explains the stability of many compounds.

Also illustrated by *Figure 2* are the low I_1 values for elements with atomic numbers 3, 11, 19 and 37. In each of these examples, loss of one electron leads to the formation of unipositive ions having the noble gas configuration.

ATOMIC SPECTRA

When atoms absorb energy, there is a possibility that their electrons will be EXCITED to shells of higher energy, the electrons emitting specific amounts of energy, known as QUANTA (or often as PHOTONS if the energy is in the form of radiation), as they return to lower energy shells. It is this energy which gives rise to emission spectra in the infrared, visible and ultraviolet regions of the electromagnetic spectrum, this emission resulting in a series of lines. These lines can be easily explained by reference to *Figure 3*, which shows the atomic spectrum of hydrogen (chosen because it is a one-electron system) and the possible transitions that give rise to this spectrum.

As can be seen from *Figure 3*, the set of lines in the higher frequency range of the spectrum results from the electron returning to the innermost energy shell, $n = 1$ (GROUND STATE), and as this represents the greatest difference in energy through which the electron can fall, these lines are also the highest in energy ($E = h\nu$, where E = energy, ν = frequency, h is Planck's constant). This series of lines is called the Lyman series. The Balmer series results from the electron returning to the second energy shell, and the Paschen series to the third. Note that the frequency of the lines may be represented by the general equation $\nu = R(1/n_1{}^2 - 1/n_2{}^2)$, where R is the Rydberg constant and n_1 and n_2 are the energy levels involved.

It can also be seen that the lines in all these series eventually converge to a limit (within each series), showing that outer shells are closer in energy to each other than are inner ones.

The Periodic Table

From atomic structure, it can now be seen that some atoms have similar electronic configurations, e.g., the noble gases have eight electrons in their outer shells (except He), and the Periodic Table arranges these similar elements

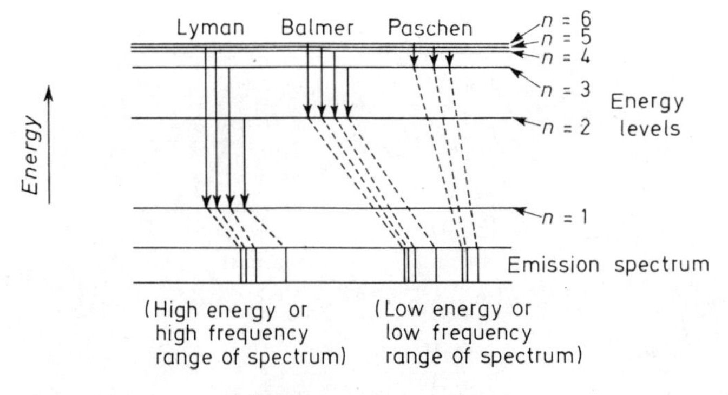

Figure 3 Emission spectrum of the hydrogen atom

together into groups or series. In the Periodic Table, each element is accompanied by its atomic number, and from these numbers electronic configurations can be written, e.g., Ne is $1s^2 2s^2 2p^6$, and hence the second shell is completed. At sodium, the third shell begins to fill with electrons (Na configuration is $1s^2 2s^2 2p^6 3s^1$ *or* [Ne] $3s^1$), and from sodium onwards electrons enter the $3s$ and $3p$ orbitals as expected up to Ar ([Ne] $3s^2 3p^6$). However, when K is reached, the nineteenth electron unexpectedly enters the fourth shell, giving K the configuration [Ar] $3d^0 4s^1$. Ca completes the $4s$ orbital, and electrons then begin to fill the $3d$ orbitals only when Sc is reached, its configuration being [Ar] $3d^1 4s^2$. From the Periodic Table, it can be seen that the elements Sc to Zn form a distinct series (called the first row transition series: *see* Chapter 13). The similarity of the elements in this series is that they all have partially filled d

orbitals either in the free atom or in one or more of their chemically important ions (except zinc), e.g., Fe is [Ar] $3d^6 4s^2$ and Ni is [Ar] $3d^8 4s^2$. However, certain configurations within the transition series are not as predicted, e.g., Cr, [Ar] $3d^5 4s^1$ and not [Ar] $3d^4 4s^2$; Cu, [Ar] $3d^{10} 4s^1$ and not [Ar] $3d^9 4s^2$. This is due to an 'exchange' of electrons between the 4*s* and 3*d* orbitals, and this occurs for two reasons:

(1) these orbitals are very near one another in energy;
(2) the filled and half-filled *d*-orbital states (d^{10} and d^5) are particularly stable configurations.

TRENDS IN GROUPS AND PERIODS

Some relationships between atomic structure and periodic properties can now be seen (*see* also Chapter 11).

(1) *Atomic Size*

The size of the atom decreases slightly across any period, because as electrons are fed into the same energy shell across any period, there is a corresponding increase in nuclear charge; consequently, this has the effect of drawing in the

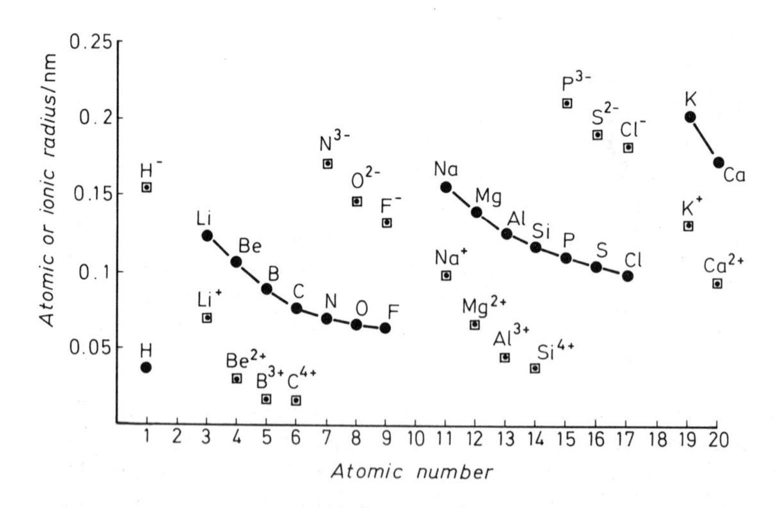

Figure 4 Relationship of atomic size to atomic number

energy shells. Reference to *Figure 4* shows this trend for elements with atomic numbers 1–20 (excluding noble gases).

As we go down a particular group, atomic size increases, because as a group is descended, each successive element has one more shell than the previous element in that group (*Figure 4*).

(2) *Ionization Energies and Electron Affinities*

Ionization energies decrease as a group is descended for two reasons:

(1) the outer shell electrons are further away from the positive nuclear charge for each successive element, and are therefore attracted less by the nucleus, and
(2) for elements near the bottom of a group, the higher atomic number means that more electrons will shield the outer shell from the influence of the nucleus (this is called electron 'screening' or 'shielding').

Generally speaking, I_1 increases across any particular period (for exceptions *see Figure 2*). This is because as the period is crossed, the nuclear charge, which is acting on the same shell, increases, and hence the electrons in this shell are more tightly held. Because of this, elements on the left-hand side of the Periodic Table have relatively low ionization energies, and in their reactions tend to lose electrons, forming positively charged ions (called CATIONS). An additional driving force for these ions to form is that the noble gas configuration is attained. Because elements on the right-hand side of a period have relatively high ionization energies, rather than lose electrons they will attempt to gain them in their reactions (forming negatively charged ions, called ANIONS). This process is often accompanied by a release of energy, known as ELECTRON AFFINITY (ΔH_e); electron affinity can, therefore, be defined as the energy involved when an isolated gaseous atom acquires an electron, e.g.:

$$Cl\cdot + e^- \rightarrow Cl^-; \ \Delta H_e = -364 \text{ kJ mol}^{-1}$$

In the case of a bivalent non-metal, the step

$$O\cdot + e^- \rightarrow O^- \text{ has } \Delta H_e = -141 \text{ kJ mol}^{-1}$$

$$\text{but } O^- + e^- \rightarrow O^{2-} \text{ has } \Delta H_e = +791 \text{ kJ mol}^{-1}$$

because energy is needed to overcome the repulsion between the O^- ion and the second incoming electron. However, energy is released again when the O^{2-} ion joins with its partner cation(s); *see* Chapter 5. (In these examples, both the ions Cl^- and O^{2-} attain the noble gas configurations of 2.8.8, and 2.8, respectively.) Note that negative energy changes indicate that energy is given out (EXOTHERMIC) whereas positive energy changes indicate that energy is taken in (ENDOTHERMIC).

Atoms in a molecule which tend to attract electrons are said to be ELECTRO-NEGATIVE (e.g., Group VII elements). Conversely, those elements which tend to release electrons (low ionization energies) such as the Group I alkali metals are said to be ELECTROPOSITIVE. As a group is descended, ionization energies decrease, so lower members of a particular group become more electropositive (or more 'metallic'). This trend is shown particularly well by the elements of Group IV, i.e., carbon is a typical non-metal, whereas at the bottom of the group, lead is a typical metal.

(It is worth noting, at this stage, that an arbitrary scale of electronegativities does exist for comparing the various elements' tendencies for attracting electrons. For example, Na = 0.9, Li = 1.0, Mg = 1.2, H = 2.1, C = 2.5, N = 3.0, O = 3.5, F = 4.0, Cl = 3.0.)

(3) *Ionic Size*

The size of the ion (*see Figure 4*) depends upon three factors:

(1) the original atomic size;
(2) the sign of the charge – positive ions are smaller than the atoms from which they originated (owing to the excess positive charge drawing in the shells), whereas negative ions are larger than the atoms from which they originated (owing to the repulsion created by the extra electrons; also, the same nuclear charge now has to act on more electrons than it did in the case of the neutral atom);
(3) the magnitude of the charge, e.g., ionic radius of $Fe^{2+} > Fe^{3+}$ and also the ionic radius of $N^{3-} > O^{2-} > F^-$

Bonding

In most chemical reactions, energy is released because the products, in which new BONDS have been formed, are in a lower energy state than the reactants. Because the noble gas configuration is stable, it is not surprising that atoms, in their reactions, attempt to achieve this structure. This can often be accomplished *via* a rearrangement of electrons during the bond formation of the new compound. However, there are examples where atoms do not achieve noble gas configuration in their compounds (*see* below).

There are two main types of bond:

(1) the COVALENT bond;
(2) the IONIC bond.

A third type, the CO-ORDINATE or DATIVE bond, is a special case of the covalent bond.

(1) COVALENT BOND

Covalent bonds are formed by atoms sharing electrons, and can therefore result in these atoms' attaining noble gas configuration. For example, in hydrogen fluoride, HF, the unpaired 2*p* electron of the fluorine atom and the unpaired 1*s* electron of the hydrogen atom share a new BONDING MOLECULAR ORBITAL (formed by the overlap of the 2*p* and 1*s* ATOMIC ORBITALS of the F and H atoms respectively) to form a single covalent bond. This situation can be represented by either of the diagrams in *Figure 5.*

Other examples of covalent compounds include water and phosphorus trichloride.

In the case of water, the two unpaired electrons in the outer shell of the oxygen atom share (two bonding molecular orbitals) with the two unpaired 1*s* electrons from two hydrogen atoms to form two O–H covalent bonds. This leaves two lone pairs of electrons in the outer shell of the oxygen atom [*Figure 6(a)*], and it is these two lone pairs which account for the shape of the water molecule (*see* Chapter 2).

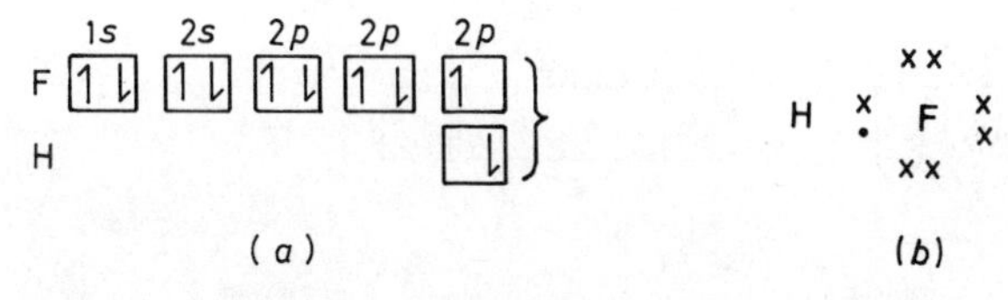

Figure 5 The two atomic orbitals shown in (a) overlap to form a new bonding molecular orbital which contains the two electrons. A conventional 'electron dot' structure for the molecule is shown in (b), where the crosses are the valency electrons of fluorine and the dot the valency electron of hydrogen

In the case of phosphorus trichloride, three unpaired electrons from the phosphorus atom share (three bonding molecular orbitals) with three unpaired electrons from three separate chlorine atoms, to form three P–Cl covalent bonds [*Figure 6*(*b*)].

Figure 6

In the above three examples, noble gas configurations are achieved by all the atoms in the compounds, but there are examples of compounds in which not all their atoms attain this structure, e.g., boron in boron trifluoride, BF_3, sulphur in sulphur hexafluoride, SF_6, and phosphorus in phosphorus pentachloride, PCl_5 (*see Figure 7*).

The ability of sulphur to form six S–F bonds in the SF_6 molecule arises from the fact that this atom possesses vacant 3*d* orbitals to which a 3*s* and a 3*p* electron can be promoted (*see Figure 8*); six unpaired electrons are now available for bonding. The energy needed for this electron promotion process (endothermic) is more than compensated for by the release of energy when the six S–F bonds are formed. In the case of oxygen, which, like sulphur, is also a Group VI element, there are no *d* orbitals in its (second) outer shell to which electrons can be promoted, and hence oxygen can never form six covalent bonds in its compounds.

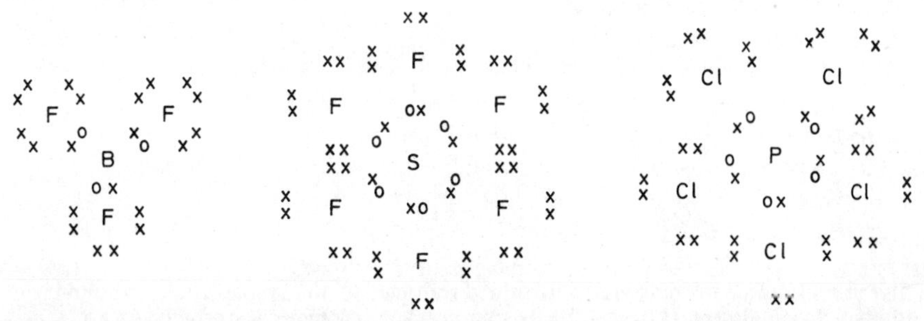

Figure 7 In these examples, instead of having either the He *structure or completed* s *and* p *orbitals only in their outer shells to give the '8-electron' noble gas structure,* B, S *and* P *have 6, 12 and 10 electrons respectively in their outer shells*

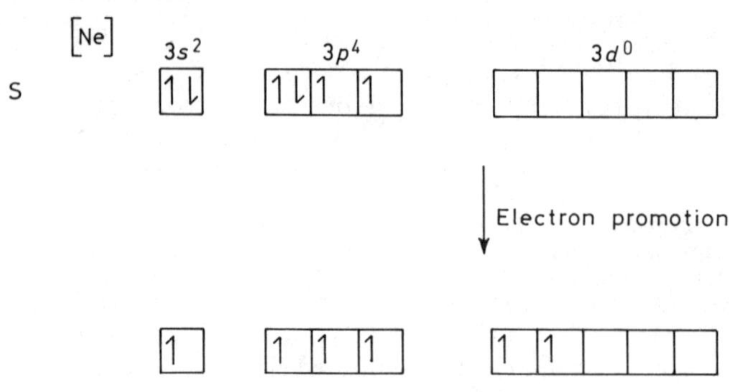

Figure 8 Six unpaired electrons are now available for sharing

Like sulphur, electron promotion can occur for carbon ($1s^2 2s^2 2p^2$); here, the 2*s* electron is promoted to the vacant 2*p* orbital resulting in four unpaired electrons being available for bonding; hence, carbon is usually quadrivalent. Again, the energy needed for electron promotion is more than compensated for by the energy released when four bonds are formed to the carbon atom. *Figure 9* shows the bond formation process for the molecule methane, CH_4.

The shape of the methane molecule is tetrahedral; it takes up this shape because the (bonding) electron pairs in the four C–H covalent bonds are then the maximum distance away from each other, hence minimizing electron repulsion. In methane the H–C–H BOND ANGLES are all 109 degrees 28 minutes, and all the BOND ENERGIES (*see* Chapter 5) are equal. In order that these four equal bonds can be formed, the four electrons in carbon's

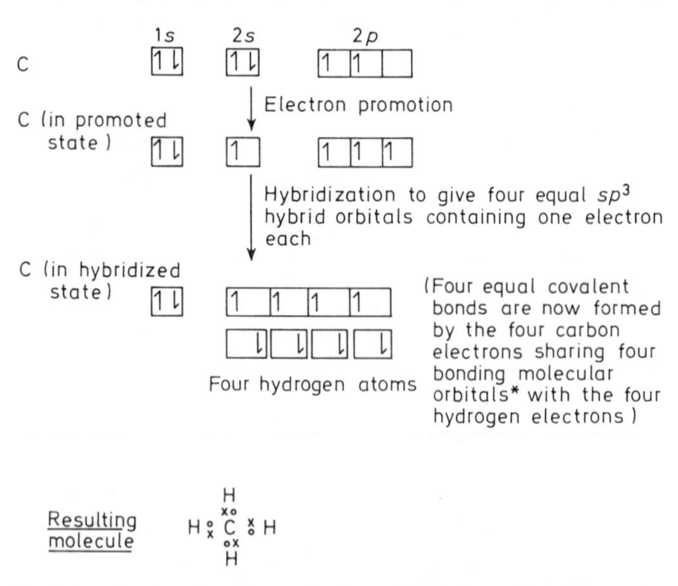

***In this molecule, a bonding molecular orbital is formed by the overlap of an sp^3 hybrid orbital with a hydrogen 1*s* orbital. Hence four new bonding molecular orbitals are made in all, each bonding molecular orbital being shared by a C electron and a H electron.**

Figure 9 Promotion and hybridization operating in the formation of the methane molecule

outer shell must occupy energetically equal orbitals, equally spaced apart. This is achieved by the 2*s* and the 2*p* orbitals 'mixing' or HYBRIDIZING to give four equal sp^3 HYBRID orbitals (*see Figure 9*); therefore, the four covalent bonds subsequently formed are all equal, both in energy and in distance apart.

A similar situation occurs for sulphur hexafluoride, where, after the electron promotion process (*Figure 8*), the 3*s*, 3*p* and two 3*d* orbitals 'mix' or hybridize to give six equal d^2sp^3 hybrid orbitals, each containing an unpaired electron. Sharing then gives six equal S–F covalent bonds (with equal bond energies and angles). The shape of this molecule is octahedral, F–S–F bond angles being 90 degrees, again minimizing electron repulsion between the (bonding) electron pairs.

Dative Bond

In the covalent bond described above, the atoms joined by the bond donated one electron each to the covalent bond. However, there are examples of covalent bonds where both electrons in the bond have been donated from the one partner atom; this is referred to as the DATIVE or CO-ORDINATE bond.

As an example, consider the formation of the ammonium ion from ammonia and a proton, H^+. Ammonia is formed by three unpaired electrons from the outer shell of the nitrogen atom sharing with three unpaired 1*s* electrons from three hydrogen atoms; this leaves a lone pair of electrons in the outer shell of the nitrogen atom, which can now form a dative (covalent) bond to the H^+ ion by donation (*Figure 10*).

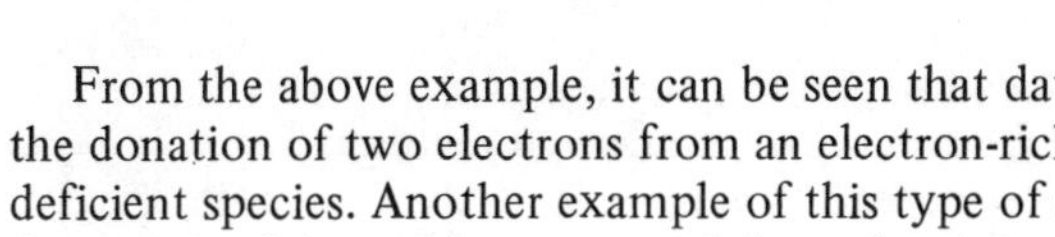

$$H_3N{:} \rightarrow H^+ \longrightarrow \left[H{-}\underset{H}{\overset{H}{N}}{-}H \right]^+$$

Figure 10 Ammonium ion, NH_4^+; *the four* N–H *bonds are all equal, differing only in origin, hence giving tetrahedral configuration*

From the above example, it can be seen that dative bonds can be formed by the donation of two electrons from an electron-rich species to an electron-deficient species. Another example of this type of covalent bond would be the formation of the stable compound from aluminium chloride and ammonia, and can be represented by any of the diagrams in *Figure 11*.

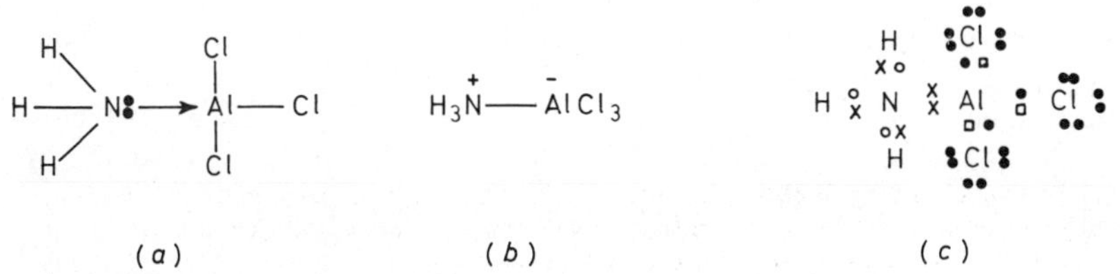

(*a*) $H_3N{:} \rightarrow AlCl_3$ (H—N: → Al—Cl with Cl above and below Al)

(*b*) $H_3\overset{+}{N}{-}\overset{-}{Al}Cl_3$

(*c*) H ° x N x x Al : Cl : (electron dot-and-cross diagram)

Figure 11

Polarization of the Covalent Bond

So far it has been assumed that electrons have been shared equally by the two atoms held by the covalent bond. However, in practice, this is seldom observed, since different atoms have different AFFINITIES for electrons. (In a molecule such as hydrogen, for example, where two identical atoms are joined by the covalent bond, then the electrons will be shared equally.) This means that in, say, the carbon to chlorine bond, the greater electronegativity of the chlorine atom results in the bonding electrons being 'pulled' towards the chlorine atom and away from the carbon atom. This effect can be represented by any of the diagrams in *Figure 12*, and the bond (or the molecule which contains the bond)

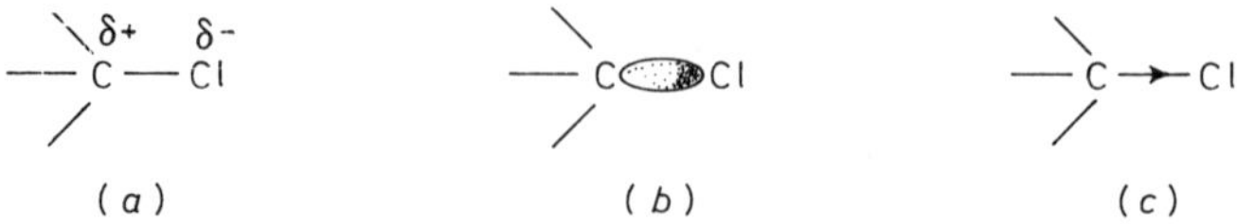

$$\overset{\delta+}{C} - \overset{\delta-}{Cl} \qquad C \text{—} Cl \qquad C \rightarrow Cl$$

(*a*) (*b*) (*c*)

Figure 12

is said to be POLARIZED. This 'pull' of electrons, in a bond, to the more electronegative atom is called the INDUCTIVE effect. There are, however, groups such as alkyl groups which can be 'electron-pushing' (*see* p. 151).

(2) IONIC (OR ELECTROVALENT) BOND

In the formation of some compounds, such as sodium chloride, where the atoms combining differ greatly in electronegativity, an electron can be lost from the more electropositive element and subsequently gained by the more electronegative element. In the case of sodium chloride, noble gas configurations are achieved:

$$\underset{(2.8.1)}{Na} + \underset{(2.8.7)}{Cl} \longrightarrow \underset{\substack{(2.8)\\ \text{Neon structure}}}{Na^+} + \underset{\substack{(2.8.8)\\ \text{Argon structure}}}{Cl^-}$$

and for calcium chloride, where a two electron transfer is involved:

$$\underset{(2.8.8.2)}{Ca} + \underset{(2.8.7\ \text{each})}{2Cl} \longrightarrow \underset{(2.8.8)}{Ca^{2+}} + \underset{(2.8.8\ \text{each})}{2Cl^-}$$

(All have argon structures)

The subsequent attraction between the newly formed positive and negative ions constitutes the IONIC BOND (N.B., noble gas structure in the ions is not always achieved, e.g., in copper(II) compounds, the Cu^{2+} ion has the electronic configuration $1s^2 2s^2 2p^6 3s^2 3p^6 3d^9 4s^0$).

The tendency to form ions, and hence ionic bonding, is greatest for elements in Groups I, II, VI and VII. Properties of ionic compounds include: existence of giant lattice structures in the solid, very often having high strengths and hence high melting points; ability to conduct electricity in their molten and, if soluble, aqueous states – this is because the ions, when free to move, act as charge carriers for conduction.

INTERMEDIATE TYPES OF BONDING

The main difference between ionic and covalent bonds is the distribution of the electrons; in the case of an ionic compound, the electron density falls to zero between the ions, whereas in covalent compounds the electron density is substantial between the two atoms joined by the bond (i.e., the electrons in a covalent bond tend to be concentrated on a line joining the two nuclei). In ionic compounds, the bond is formed by the attraction of the oppositely charged ions, whereas in a covalent compound the bond is formed by the shared electron pair 'cloud' (between the nuclei) attracting the nuclei to it and hence towards each other.

The question now arises whether bonds are purely ionic, purely covalent, or whether an intermediate type of bond exists. The answer is that intermediate types of bond do occur, arising from two processes. The first, polarization of covalent bonds, has already been discussed, and it can be said that this process introduces some 'ionic character' into the bond. The second process is the polarization of ions (*Figure 13*) where the electron 'cloud' of one or both of the

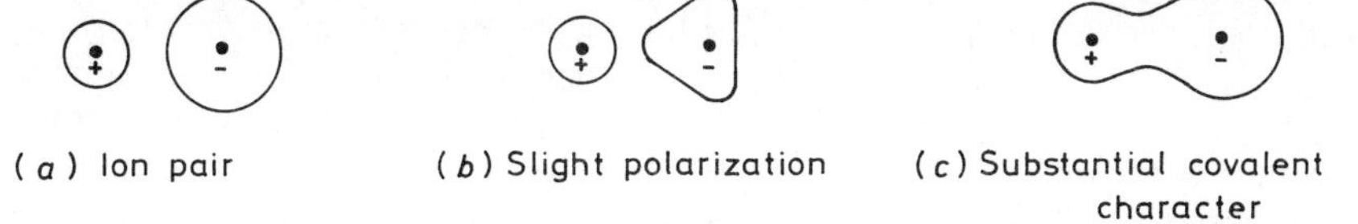

Figure 13

ions becomes distorted. This second process introduces some 'covalent character' into the ionic bond, and the extent to which this happens can be summarized by Fajans's rules; these state that a compound will have appreciable covalent character if:

(1) either the cation or anion is highly charged (making the cation polarizing and the anion polarizable);
(2) the cation is small (hence positive charge density over 'surface' of cation is high and will therefore polarize the anion);
(3) the anion is large (outer electrons are far from the nucleus and so less under its influence; therefore, distortion of the anion's electron 'cloud' is easier).

Consequently, wholly ionic and wholly covalent bonds are extreme types and, in practice, bonds can be partially ionic and partially covalent in character.

EFFECT OF BONDING ON SOLUBILITIES

When discussing solubilities and solutions, it is useful to label the material present in excess as the SOLVENT, and that in the lower concentration as the SOLUTE. In predicting the solubility of a solute in a particular solvent, the bonding or degree of polarization of both solute and solvent has to be considered, since it is found experimentally that many ionic or polar solutes are soluble in polar solvents, whereas non-polar solutes tend to be soluble in non-polar solvents, e.g., salt is soluble in water (a polar solvent) but iodine (which is a non-polar solute) is found to be virtually insoluble in water but soluble in benzene (a non-polar solvent). The reasons for this are outlined in Chapter 3.

1

NATURE OF ATOM AND RADIOACTIVITY

1.1 The Atom

The total mass of the atom is largely made up of the mass of the nucleus (*see Table 1*), which in turn depends on the number of neutrons and protons present. Atoms of the same element, although having the same number of protons and electrons, may differ in mass because they can have different numbers of neutrons in their nuclei. These different forms of the same element are called ISOTOPES, e.g., hydrogen has three isotopic forms; hydrogen (1 electron, 1 proton), deuterium (1 electron, 1 proton and 1 neutron) and tritium (1 electron, 1 proton and 2 neutrons). They are written as: ${}^{1}_{1}H$, ${}^{2}_{1}H$ and ${}^{3}_{1}H$ respectively. [N.B., the superscript or MASS NUMBER (no. of protons + no. of neutrons) is the mass of the isotope to the nearest whole number.] The last has an unstable nucleus and is said to be RADIOACTIVE.

In determining RELATIVE ATOMIC MASSES (atomic weights), the current convention is to define the isotope ${}^{12}_{6}C$ as 'weighing' 12.00000 units, and all other elements in their natural states are compared with this standard. Most elements do not have relative atomic masses which are integers, because most elements in their natural states consist of a mixture of their isotopes, e.g., chlorine occurs as ${}^{35}_{17}Cl$ and ${}^{37}_{17}Cl$. Also, electrons do have a small mass, and protons and neutrons do not have *exactly* the same mass.

Because of isotopes, different elements can have the same total number of protons and neutrons (and therefore the same mass number); such isotopes are called ISOBARS, e.g., ${}^{40}_{18}Ar$, ${}^{40}_{19}K$ and ${}^{40}_{20}Ca$ form one set of isobars.

1.2 Fundamental Particles

1.2.1 ELECTRON

Thomson found that if an electric discharge were passed through a gas at low pressure, a beam of rays was emitted from the negative electrode; if passed through an electric field, the rays were deflected towards the positive pole, showing they were negatively charged. Thomson also found that the mass to charge ratios (m/e) for these rays were constant irrespective of the gas used, indicating the particles to be uniform (i.e., the electron). The apparatus used for measuring m/e values is called the mass spectrometer, and is now used extensively for measuring m/e values of cations in chemical analysis (*see* below).

Millikan determined the charge on the electron using a method which depended on charging fine oil drops with one or more electrons, and allowing them to fall between two plates connected to a variable high voltage supply (*see Figure 1.1*). By making the top plate positive, and by altering the strength

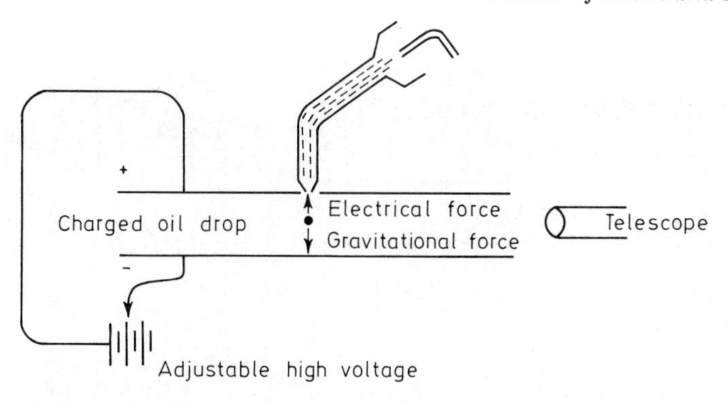

Figure 1.1 Millikan's oil-drop experiment

of the electric field until the drop remained stationary, the electrical force and hence the charge on the drop could be calculated. The smallest charge found on any drop was 1.602×10^{-19} C; furthermore, all other charges were found to be multiples of this charge. Millikan therefore assumed, reasonably, that this was the fundamental electric charge, and also the charge on the electron. The mass of the electron could now be obtained from the charge and m/e values.

1.2.2 PROTON

Rutherford studied the deflection of a beam of α-particles (positively charged particles, *see* below) projected at a very thin piece of metal foil; he found that although most of the α-particles passed through the foil with little or no deviation, some were deflected strongly. Hence he deduced that atoms in the foil must have regions of high positive charge density (but that most of the atom was empty space).

Electric discharges in gases produce both electrons and positively charged atomic or molecular fragments, the latter being produced by ionization. Using the mass spectrometer, m/e values can be determined for these cations. The extent of deflection of the cation in an electric or magnetic field depends on (1) charge and (2) velocity of cation, which in turn depends on mass (the lower the velocity, the more extreme is the deflection). A simple mass spectrometer is shown in *Figure 1.2*.

Each cation, therefore, has its own characteristic path in the field, dependent on its mass and charge. The smallest m/e value obtained is when hydrogen is the sample material. It is therefore assumed that the hydrogen nucleus contains a single positive particle, the proton.

The mass spectrometer is currently used as a tool for chemical analysis. The unknown vaporized sample is ionized and/or fragmented by a beam of electrons as above. The m/e values (relative to proton = 1) of the ionized molecules and fragments can then give information concerning the relative molecular mass and structure of the unknown (and also of the various isotopes present).

1.2.3 NEUTRON

The existence of neutrons was suggested as a result of experiments done by Rutherford and others (*see* before). Rutherford found that he could determine

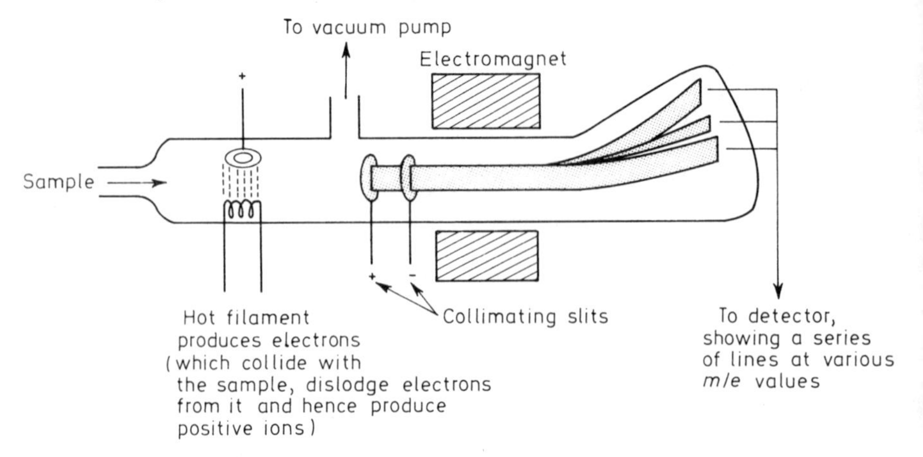

Figure 1.2 The mass spectrometer

the charge on a nucleus from the pattern made by deflected α-particles, since the greater the charge on the nucleus the greater the deflections. He then observed that there was no correlation between the actual mass of an atom (from, say, mass spectrometer information) and the total mass of the protons and electrons present; he therefore suggested the existence of a neutral particle in the nucleus. Chadwick later found that 'radiation' emitted when beryllium was irradiated with α-particles behaved as neutral particles, i.e., neutrons.

1.2.4 WAVE AND PARTICLE BEHAVIOUR OF ELECTRON

de Broglie suggested that if Planck's equation, $E = h\nu$ (*see* Introduction), and Einstein's equation $E = mc^2$ (which relates energy E to mass m, where c is velocity of light) were applicable to electrons, then on combination, $h\nu = mc^2$ or $h = mc^2/\nu$. Since $c = \nu\lambda$ (where λ = wavelength), then $\nu = c/\lambda$ or, on substitution, $h = mc\lambda$ or $\lambda = h/mc$.

The properties of a wave and a particle have been combined. The electron does, in reality, show both the properties of a particle (the determination of its mass, for example, has just been described) and of a wave (electrons do undergo diffraction and interference effects; *see* Chapter 4). The electron, consequently, is said to have a DUALITY of particle and wave character.

The Heisenberg uncertainty principle says that the exact position and velocity of an electron cannot be known at a given time. Instead, the wave mechanical (or quantum mechanical) treatment of this problem (based on an equation called the Schrödinger wave equation) gives the PROBABILITY of finding an electron at a given position. Consequently, the 1*s* orbital can be described as a region in space where there is a MAXIMUM PROBABILITY of finding the 1*s* electron in, say, the hydrogen atom.

1.3 Radioactivity

Becquerel first observed one of the effects of radioactivity; he noted that a uranium salt blackened a photographic plate with which it was in contact, even when the plate was wrapped in light-proof paper.

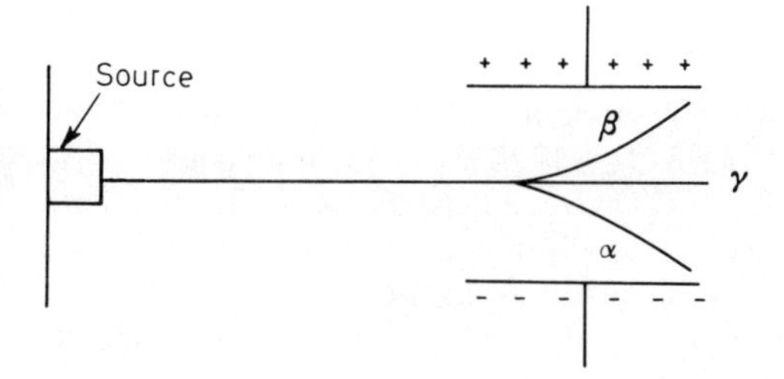

Figure 1.3 Deflection of α, β and γ 'rays' in an electric field

It was Rutherford who first showed that radiation could be divided into three main types; alpha (α), beta (β) and gamma (γ). *Figure 1.3* shows the behaviour of α, β and γ-rays when passed between two oppositely charged plates. The following should be noted:

(1) α-Rays are fast moving, positively charged particles having a mass of 4 and a charge of +2, i.e., helium nuclei; α-particles are readily absorbed by about 7 cm of air. They are affected by a magnetic field.
(2) β-Rays are fast moving electrons; they will penetrate thin aluminium but are stopped by lead. They are greatly affected bv a magnetic field.
(3) γ-Rays are electromagnetic waves (i.e., no charge), and behave similarly to short wavelength *X*-rays (*Figure 1.4*) but are more penetrating. (*X*-Rays are emitted when inner atomic electrons near a nucleus return to the ground state after having been excited.) γ-Rays can penetrate metal, including lead.

Properties of radioactivity (some of which can be used for detection of radiation) include: action on a photographic plate; the ionization of gases (which is used as the basis of the Geiger–Müller counter); will harm living cells; can cause fluorescence in some compounds, e.g., zinc sulphide (fluorescence is the emission of radiation caused by excited electrons returning to the ground

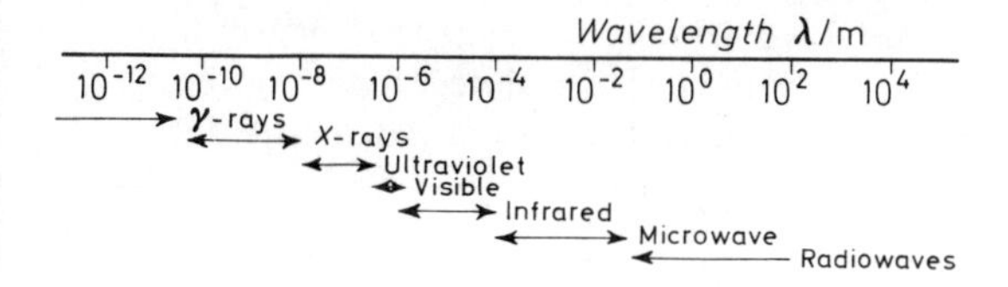

Figure 1.4 The electromagnetic spectrum

state, after having first been irradiated; fluorescence ceases virtually when irradiation ceases).

1.3.1 HALF-LIFE OF RADIOACTIVE ISOTOPES

The half-life ($t_{1/2}$) of a radioactive isotope is defined as the time taken for half the number of unstable nuclei present to disintegrate. The half-life is characteristic of a particular radioactive isotope.

1.3.2 REASONS FOR NUCLEAR INSTABILITY

The $^{16}_{8}O$ isotope contains eight protons, eight neutrons and eight electrons. The sum of the masses of these 24 'particles' is 16.13200 amu. Accurate mass measurement (by mass spectrometer) of $^{16}_{8}O$ gives a mass of 15.99491 amu, i.e., a mass difference of 0.13709 amu. This mass difference is called the MASS DEFECT and is a measure of the energy needed to split the atom into its constituent particles (energy is related to mass by $E = mc^2$, *see* before). The energy which appears to have been released when the $^{16}_{8}O$ atom is formed from its fundamental particles is called the BINDING ENERGY of the atom. Data (such as the above) show that stable isotopes have larger mass defects and binding energies than do radioactive (unstable) ones.

1.3.3 EXAMPLES OF RADIOACTIVE DECAY

Table 1.1 gives the symbols for some of the fundamental 'particles' in radiochemistry theory. All isotopes with atomic numbers greater than 83 are radioactive. When α-particles are emitted, the nucleus loses two protons and two neutrons, i.e., the atomic number decreases by 2 and the mass number by 4.

Table 1.1 Fundamental 'particles' for radioactivity

Emission	*Symbol*	*Emission*	*Symbol*
α-Particle	$^{4}_{2}He$	Proton	$^{1}_{1}H$
β-Particle	$^{0}_{-1}e$	Neutron	$^{1}_{0}n$
Positron*	$^{0}_{+1}e$	γ-Ray	γ

*A sub-atomic particle identified in disintegration processes, which behaves like a 'positive electron'

Alternatively, if a β-particle (electron) is emitted from the nucleus, the atomic number increases by 1 but the mass stays virtually the same (and hence the mass number), since a neutron can be thought of as disintegrating to give an electron and a proton. The following are some examples of radioactive decay:

$$^{238}_{92}U \rightarrow {}^{4}_{2}He + {}^{234}_{90}Th \qquad (\alpha\text{-emission})$$

$$^{234}_{90}Th \rightarrow {}^{0}_{-1}e + {}^{234}_{91}Pa \qquad (\beta\text{-emission})$$

$$^{234}_{91}Pa \rightarrow {}^{0}_{-1}e + {}^{234}_{92}U \qquad (\beta\text{-emission})$$

Note that there is a balance of nuclear particles and charge, i.e., sum of superscripts of reactants = sum of those of products, and similarly for the subscripts.

If the number of neutrons is plotted against the number of protons for the stable isotopes of the elements, the points of the graph lie on a curve called the 'stability band' (*Figure 1.5*); consequently, isotopes not lying in this band will decay until they do so. Nuclei which lie in region *A* (*Figure 1.5*) are neutron rich, and achieve stability by β-emission. This can be pictured as the transformation of a neutron into a proton and an electron ($^{1}_{0}n \rightarrow {}^{0}_{-1}e + {}^{1}_{1}H$), hence bringing the resulting nucleus nearer the stability band. In contrast, nuclei in region *B*

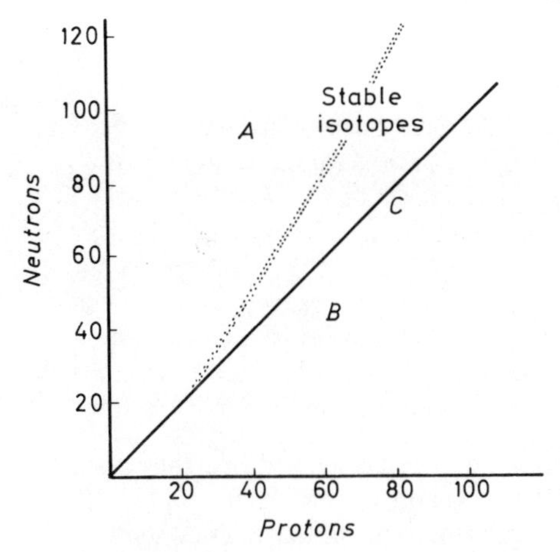

Figure 1.5 Stability of isotopes. Region A contains neutron-rich and region B proton-rich isotopes. Line C corresponds to a neutron:proton ratio of 1:1

(neutron deficient) can achieve stability by positron emission, since this results in a proton being converted into a neutron ($^1_1H \rightarrow {}^1_0n + {}^{\;0}_{+1}e$). Nuclei in region *B* can also undergo *K*-capture, which involves the capture of an orbital electron (from the first shell), hence converting a proton into a neutron.

It should be noted that in any disintegration process, the rate of disintegration of any isotope is directly proportional to the number of radioactive nuclei present. This process is called a FIRST-ORDER process (*see* Chapter 10).

1.3.4 TRANSMUTATION REACTIONS

Reactions between nuclei or between nuclei and other particles can often lead to the synthesis of new nuclei; this is called TRANSMUTATION. For example, in the collision of high-energy protons with lithium, α-particles are produced:

$$^7_3Li + {}^1_1H \rightarrow 2{}^4_2He$$

1.3.5 NUCLEAR FISSION

When the $^{235}_{92}U$ isotope captures a neutron, the resulting nucleus can disintegrate into fragments of roughly equal size; this process is called nuclear fission:

$$^{235}_{92}U + {}^1_0n \rightarrow {}^{95}_{42}Mo + {}^{139}_{57}La + 2{}^1_0n + 7{}^{\;0}_{-1}e$$

1.3.6 NUCLEAR FUSION

This is the fusion of light nuclei accompanied by the release of vast quantities of energy, e.g.,

$$^2_1H + {}^3_1H \rightarrow {}^4_2He + {}^1_0n$$

(Solar energy is generated by fusion processes.)

2

SHAPES OF MOLECULES

2.1 Electron Pair Repulsion

The shape of the methane molecule has already been discussed (*see* Introduction). Shapes of other molecules can also be explained using electron-pair repulsion theory; here, it is assumed that electron pairs in the valency shells of the atoms will repel one another and so try to get as far apart as possible. This situation is summarized in *Figures 2.1* and *2.2* where shapes and bond angles, together with the number of valency electron pairs (given in parentheses), are given for some common molecules and ions.

In the case, for example, of methane, all pairs of valency electrons are equivalent. For ammonia, however [*Figure 2.2(a)*], the four pairs of valency

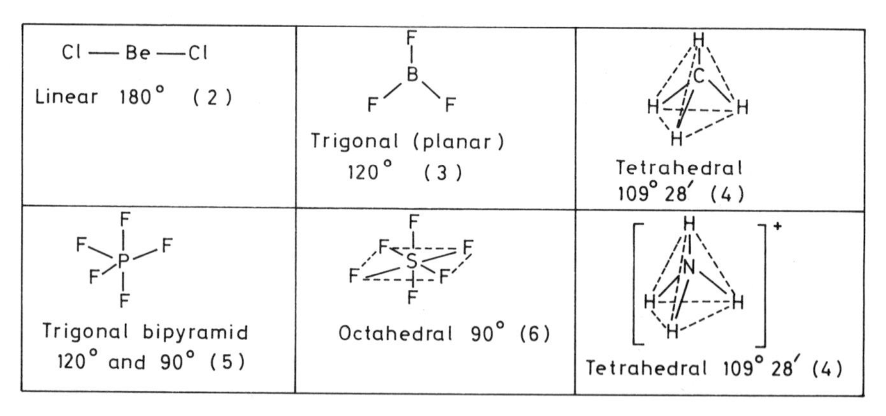

Figure 2.1 Shapes and bond angles of some simple molecules and ions

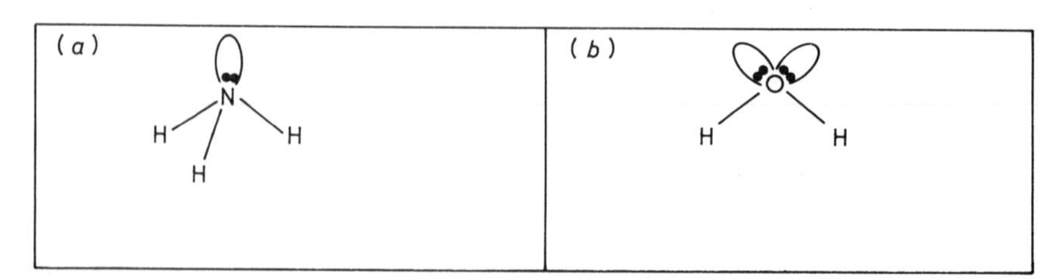

Figure 2.2 Tetrahedral shapes due to lone pairs. In the ammonia molecule (a), the structure is pyramidal, based on a tetrahedron, the H–N–H *angle being* 107.5 *degrees. In the water molecule (b), the two lone pairs and the two bonding pairs of electrons make a distorted tetrahedral set, the* H–O–H *angle being* 104.5 *degrees*

electrons are not equivalent, there being three bonding pairs and one non-bonding pair (lone pair); consequently, the H–N–H bond angles are distorted slightly from the tetrahedral angles of 109.5 degrees. In the case of the water molecule [*Figure 2.2*(*b*)] there are two non-bonding pairs (or lone pairs) of electrons, and so distortion from the tetrahedral shape is greater*. Other factors which can distort a molecule from the basic shape are (1) if one (or more) of the valency electron pairs is bonded to a different atom (as occurs in, say, chloromethane, CH_3Cl) since this makes the bond lengths different, and (2) if electron pairs are involved in multiple bonds (*see* later).

Figure 2.3 shows how shapes can be predicted by extending these ideas (shapes of transition metal complexes are discussed in Chapter 13).

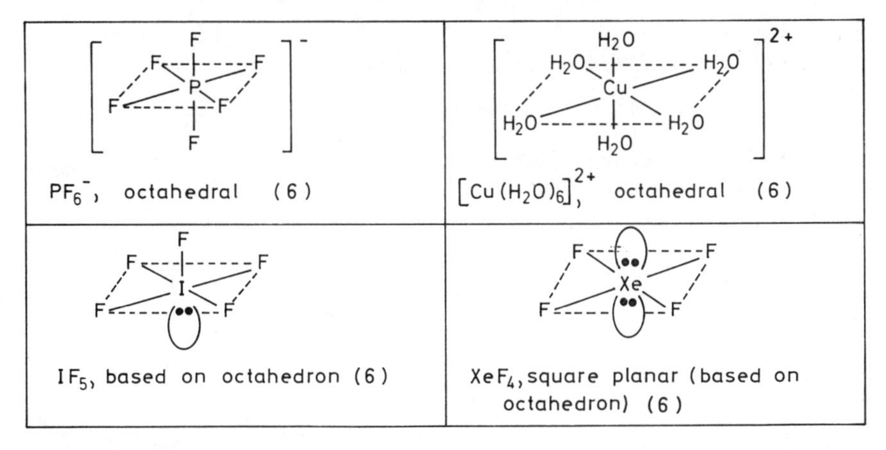

Figure 2.3 Octahedral shapes

Note how chemical reaction can influence shape. For example, the shape of BF_3 is trigonal (planar); when ammonia is added to form $H_3N \rightarrow BF_3$, the fluorine atoms take up a roughly tetrahedral configuration round the boron as the nitrogen atom is drawn in.

Shapes of organic molecules can also be predicted fairly accurately by considering the number of atoms or groups attached to the carbon atom under consideration. For example, methane is tetrahedral: the carbon atomic orbitals involved are sp^3 hybrid orbitals (*see* Introduction). Similarly, whenever a carbon atom is saturated (i.e., carbon forms four single covalent bonds to other atoms) the approximately tetrahedral configuration is observed, e.g., in ethane, C_2H_6, where all bond angles are roughly the tetrahedral angle.

The bonding in ethene, C_2H_4, however, involves different hybrid orbitals, called sp^2 hybrid orbitals; their formation is shown in *Figure 2.4* (together with the orbital overlap responsible for subsequent covalent bond formation in ethene). An sp^2 carbon is referred to as a trigonal carbon atom, with approximately 120 degree angles between the bonds to attached atoms and hence giving a planar shape. Ethene is therefore a planar molecule. Similarly, parts of organic molecules containing a carbon bonded to only three atoms (i.e., sp^2 carbon),

*Summarizing, lone pair–lone pair repulsion > lone pair–bonding pair repulsion > bonding pair–bonding pair repulsion

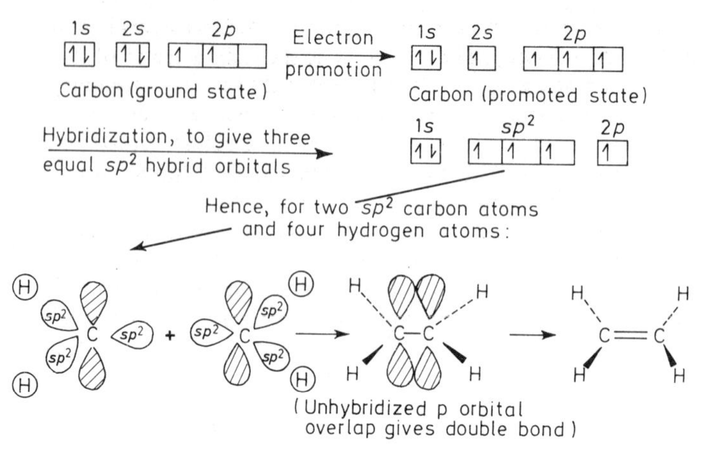

Figure 2.4 Hybridization of carbon leading to planar ethene

such as in carbonyl compounds (>C=O group, *see* Chapter 15, p. 170) also have this planar structure. In ethanal, CH_3CHO, for example, the $-CH_3$ part of the molecule would have a tetrahedral configuration, and the –CHO part would be planar.

For ethyne, C_2H_2, *sp* hybridization of carbon atoms is involved. The subsequent sharing of electrons (because of orbital overlap) results in a linear molecule being formed, bond angles 180 degrees.

1s	2s	2p		1s	sp	2p
↑↓	↑	↑ ↑ ↑	Hybridization, → to give two equal *sp* hybrid orbitals	↑↓	↑ ↑	↑ ↑

(Carbon in promoted state)

It should be noted that the C–C bond lengths in ethane, ethene and ethyne are 0.154, 0.134 and 0.121 nm respectively, showing how multiple bonds cause interatomic distances to contract.

2.2 The Effect of Resonance on Molecular Shape

Experimentally, it is found that the C–C bond lengths in benzene, C_6H_6, are all 0.139 nm showing that they are intermediate in character between double and single C–C bonds. As expected from *Figure 2.5(a)*, each carbon atom is sp^2 hybridized (hence the molecule is planar, bond angles 120 degrees). However,

usually written as

(*a*) (*b*) (*c*)

Figure 2.5 Delocalization in benzene

the six remaining $2p$ electrons on the six carbon atoms (left after hybridization; *see* ethene) are said to be DELOCALIZED over these six carbon atoms of the ring, because of the possibility of any particular carbon $2p$ orbital overlapping with $2p$ orbitals on adjacent carbon atoms. Consequently, benzene can be regarded as a RESONANCE HYBRID of the two Kekulé structures shown in *Figure 2.5(b)*. This does not mean that benzene is continually 'flipping' from one structure to the other, but that the true structure is somewhere 'in between' these two forms (hence explaining the intermediate bond length values). This delocalization of electrons in the benzene ring is represented by the structure in *Figure 2.5(c)*. The process by which bonds are rearranged in a molecule is termed RESONANCE or MESOMERISM, and the more resonance structures which can be drawn for a particular molecule, the more stable it is (i.e., delocalization of electrons stabilizes molecules); this accounts for the stability of benzene (*see* Chapters 5 and 15).

Other examples of resonance can be seen; electron diffraction studies show that in nitric acid, the two N–O bonds (not the N–OH bond) are equal in length, because of resonance.

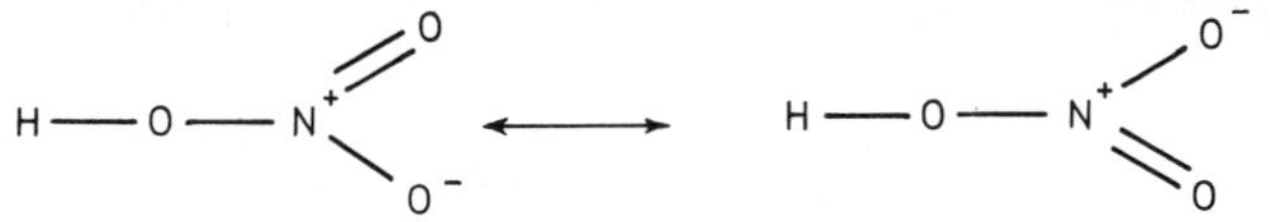

$$H{-}O{-}N^{+}({=}O)({-}O^{-}) \longleftrightarrow H{-}O{-}N^{+}({-}O^{-})({=}O)$$

Resonance also occurs in the nitrate, sulphate and carbonate ions:

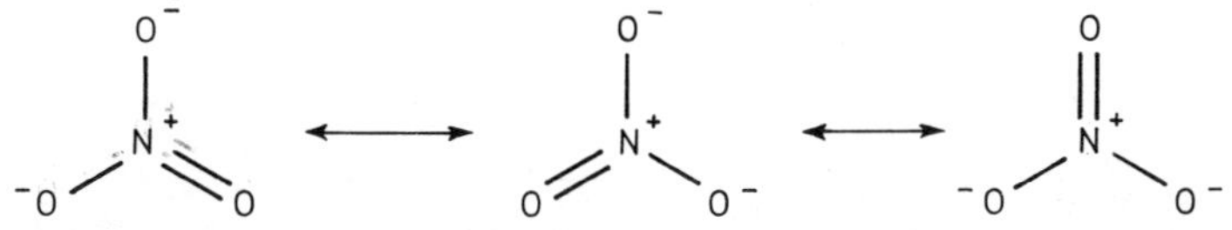

$$N^{+}(O^{-})(O^{-})({=}O) \longleftrightarrow N^{+}(O^{-})({=}O)(O^{-}) \longleftrightarrow N^{+}({=}O)(O^{-})(O^{-})$$

(Planar, bond angles 120 degrees, all N–O bond lengths equal)

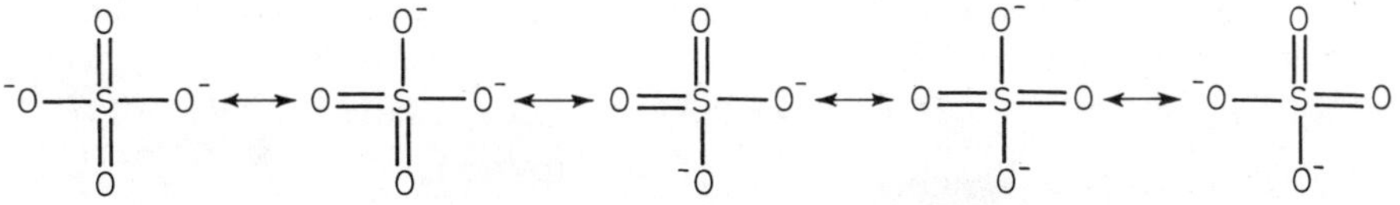

$$^{-}O{-}S({=}O)_2{-}O^{-} \longleftrightarrow O{=}S(O^{-})({=}O){-}O^{-} \longleftrightarrow O{=}S({=}O)(O^{-}){-}O^{-} \longleftrightarrow O{=}S(O^{-})(O^{-}){=}O \longleftrightarrow {}^{-}O{-}S({=}O)(O^{-}){=}O$$

(All S–O bonds have same length, bond angles are tetrahedral)

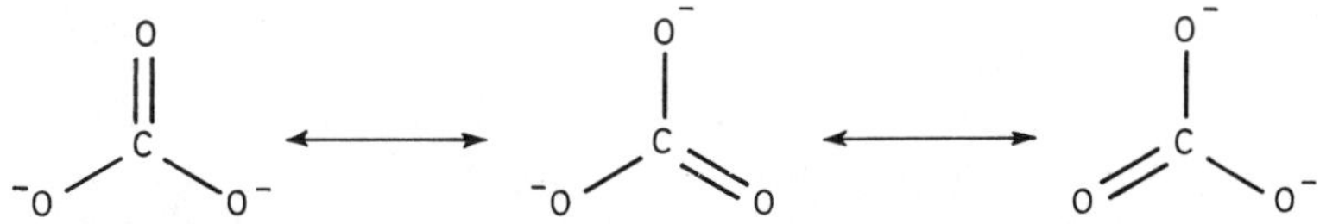

$$C({=}O)(O^{-})(O^{-}) \longleftrightarrow C(O^{-})(O^{-})({=}O) \longleftrightarrow C(O^{-})({=}O)(O^{-})$$

(Planar, bond angles 120 degrees, all C–O bond lengths equal)

3

FORCES BETWEEN ATOMS, MOLECULES AND IONS

3.1 Intermolecular and Intramolecular Forces

Several different types of force exist between atoms, molecules and ions. Those important in the solid state include covalently bonded two- and three-dimensional networks (e.g., graphite and diamond), ionic forces (e.g., sodium chloride crystal) and metallic bonding; these are discussed in Chapter 4. Other types of force include van der Waals forces, dipole–dipole attractions and hydrogen bonding; these can be found in all three states of matter since they are of varying magnitude.

In a covalent molecule, the electrons within that bond will not be shared equally if the atoms joined by the bond have different electronegativities. This leads to polarization of the bond (*see* Introduction). If this difference in electronegativity is large, then the bond will be strongly polarized, but if the atoms have similar electronegativities, the bond will be only slightly polarized. In the

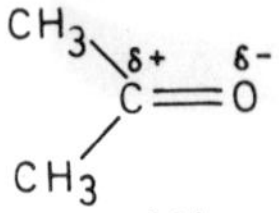

$$\begin{array}{l} CH_3 \\ \quad \searrow^{\delta+} \quad ^{\delta-} \\ \quad\; C{=}O \\ \quad / \\ CH_3 \end{array}$$

(1)

case of propanone (acetone), for example, the molecule is permanently polarized; the delta (δ) signs in structure (1) indicate that the carbon atom has a fractional positive charge and the oxygen atom has a fractional negative charge. This charge separation is expressed and measured as the dipole moment (a dipole is a separation of two opposite charges by a small distance). DIPOLE–DIPOLE attractions now occur between molecules, i.e., the $\delta-$ region of one molecule will therefore attract the $\delta+$ region of an adjacent molecule. Propanone, therefore, has a boiling point of 56 °C which is high compared with that of butane (b.p. 0 °C), a non-polar molecule having the same relative molecular mass. It should be appreciated that in the case of a molecule like carbon dioxide, O=C=O, there is no permanent dipole since the inductive effects are equal and opposite because of the molecule's symmetry. However, there are still regions of high and low electron density (oxygen atoms are $\delta-$ and the carbon atom $\delta+$) so there will still be electrostatic forces operating between the carbon dioxide molecules. (Any type of force which operates between molecules is referred to as an INTERMOLECULAR force.)

VAN DER WAALS forces are thought to be due to continually changing dipole–induced dipole interactions between atoms or molecules. The 'electron-cloud' of an atom or molecule is in continual motion, and can therefore be more on one side of the atom (or molecule) than the other at any given time; consequently, a flickering dipole is set up. This dipole can now induce a dipole on a neighbouring atom or molecule, and a force of attraction results. These van der Waals forces are very weak. They increase in size as the number of electrons in the atom or molecule increases. For example, the boiling points of the halogens increase as Group VII is descended, and this is not just because of the increase in relative molecular mass; the larger the molecule, the more electrons it contains and hence the more polarizable it becomes (because outer or valency electrons are less under the control of the nucleus – *see* Introduction). Very low temperatures are needed to liquefy the noble gases, but the fact that they do liquefy does indicate the presence of some weak interatomic forces – these are, in fact, van der Waals forces.

HYDROGEN BONDING is a specific type of dipole–dipole interaction and can occur between molecules which contain a hydrogen atom (or atoms) covalently bonded to a highly electronegative atom. The hydrogen atom has only one electron, and in the covalent bond this will be nearer the electronegative atom than the hydrogen atom. Thus, a large part of the hydrogen nucleus is devoid of screening electrons, and will therefore be attracted to regions of high electron density such as lone pairs on oxygen, nitrogen and fluorine atoms. Water is a liquid at room temperature, whilst hydrogen sulphide is a gas (*see Table 3.1*); this is a direct consequence of hydrogen bonding between water molecules [*Figure 3.1(a)*] because of the higher electronegativity of oxygen compared with that of sulphur. In the case of ice, a loosely packed (three dimensional) tetrahedral structure is observed [*Figure 3.1(b)*]. When ice melts, the

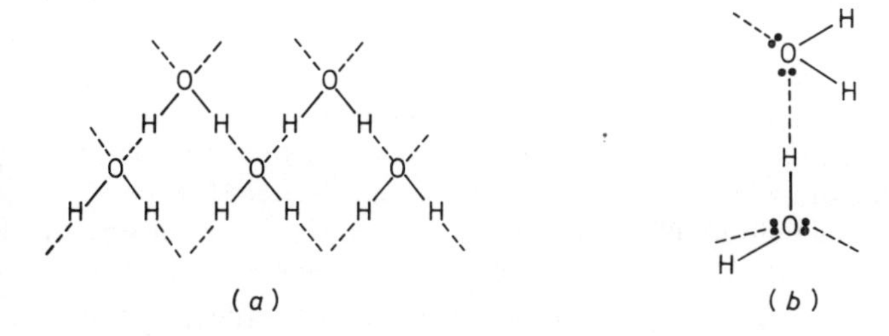

Figure 3.1 Hydrogen bonding; the dotted lines indicate hydrogen bonds

density of water increases to a maximum value at 4 °C because the ice structure collapses and closer packing of hydrogen bonded units can occur – above 4 °C, the density decreases because of normal thermal effects. Another important example of hydrogen bonding occurs in proteins between the amide or peptide groups on the same molecule (called INTRAMOLECULAR bonding).

Table 3.1 gives a comparison of the boiling points of some hydrides of the second and third periods. It can be seen, from these boiling points, that intermolecular attractions increase (owing to hydrogen bonding) as the hydrogen

Table 3.1 Boiling points of hydrides

2nd Period hydride	*B.p./°C*	*3rd Period hydride*	*B.p./°C*
CH_4	−161	SiH_4	−112
NH_3	−33	PH_3	−90
H_2O	+100	H_2S	−60
HF	+20	HCl	−85

atom is attached to atoms of increasing electronegativity. Note that in the case of HF, despite the fact that fluorine is the most electronegative element, there is only one hydrogen atom per molecule to hydrogen bond to an adjacent HF molecule. In the case of methane and silane, where C and Si are not electronegative elements, hydrogen bonding does not occur; the increased boiling point of SiH_4 is because of the larger relative molecular mass and the larger van der Waals forces which operate between molecules.

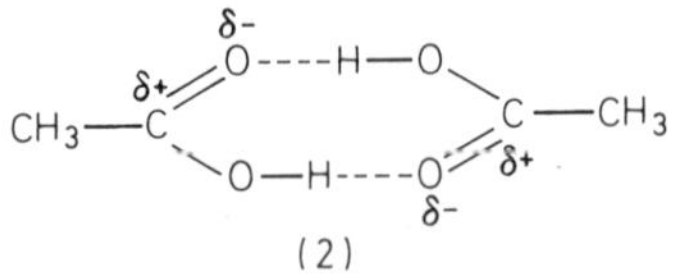

(2)

Other evidence for hydrogen bonding is shown by the DIMERIZATION of some acids. For example, when ethanoic acid is dissolved in a non-polar solvent (e.g., benzene) the relative molecular mass is found to be 120, not 60 – ASSOCIATION has occurred because of hydrogen bonding, as shown in structure (2). In general, the hydrogen bond is a strong form of dipole–dipole interaction, and is intermediate in strength between van der Waals forces and covalent bonds.

3.2 Solvation

As stated in the Introduction, ionic or polar solutes tend to be soluble in polar solvents, whereas non-polar solutes tend to be soluble in non-polar solvents. In ionic solids, the giant lattice structure is strongly held together by the powerful electrostatic forces operating between the positive and negative ions (hence giving rise to high melting points). When an ionic solid dissolves in a polar solvent the crystal lattice breaks up or DISSOCIATES, which often requires a large amount of energy (endothermic – *see* lattice energy, p. 44). However, energy is released again (exothermic) when the separated ions become SOLVATED, i.e., new links or bonds are produced between the ions and the solvent molecules (either by ion–dipole attractions (*Figure 3.2*) or by co-ordinate bond formation – *see* Chapter 13). If the solvent is water, the process of solvation is known as HYDRATION. This process is shown in *Figure 3.2* for sodium chloride.

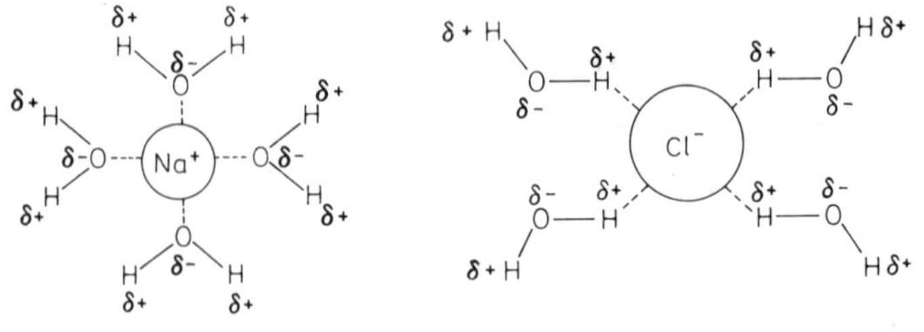

Figure 3.2 Solvation of ions by water

Many polar molecules are found to dissolve in polar solvents, e.g., propanone (acetone) dissolves in water. Propylamine, $C_3H_7NH_2$ [*Figure 3.3(a)*], also dissolves in water; here, not only are there dipole–dipole attractions but also hydrogen bonding can occur [*Figure 3.3(b)*].

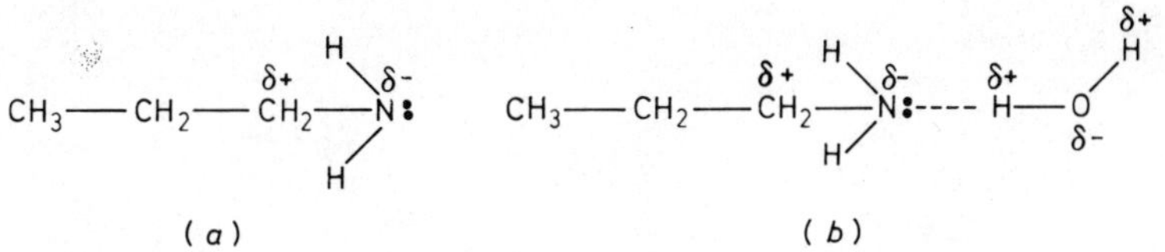

$$\underset{}{CH_3}—CH_2—\overset{\delta+}{CH_2}—\overset{\delta-}{N}H_2\text{:}$$

(*a*)

$$CH_3—CH_2—\overset{\delta+}{CH_2}—\overset{\delta-}{N}H_2\text{:}\cdots\overset{\delta+}{H}—\underset{\delta-}{O}—\overset{\delta+}{H}$$

(*b*)

Figure 3.3 Dipole–dipole attractions

The general trend is that solubility can occur when the solute–solvent interactions are similar to or greater in size than the solute–solute and solvent–solvent interactions. For example, in the case of a non-polar solute such as iodine, iodine molecules (formed by two covalently bonded iodine atoms) are held to each other in the solid by weak van der Waals forces; benzene molecules are held to each other by the same type of force. These are similar in magnitude to the iodine–benzene van der Waals forces and hence benzene can easily penetrate the iodine crystal, separating the iodine molecules, and solution occurs. However, iodine is only sparingly soluble in water because the water–water interactions (i.e., hydrogen bonds) are stronger than the iodine–iodine and iodine–water interactions.

Another factor, namely ENTROPY, also affects solubility, and this is discussed in Chapter 5.

4

STATES OF MATTER

4.1 Introduction

In the gaseous state, molecules (or atoms in the case of the noble gases) move about in a state of random, rapid motion. The molecules are far apart and have little influence upon each other except during collisions. If the temperature of the gas is lowered, the kinetic energy, and hence velocity, of the molecules is lowered. If, also, the pressure on the gas is increased, the molecules are forced closer together (causing intermolecular attractions to increase). Hence, lowering the temperature or increasing the pressure can cause liquefaction, and this change is called a FIRST-ORDER TRANSITION resulting in a PHASE CHANGE from gas to liquid. If the temperature is lowered further, more and more of the molecules interact until virtually every molecule is strongly attracted by neighbouring molecules – another first-order transition occurs with the phase change from liquid to solid (known as the freezing or melting point). Molecules in liquids are therefore in an intermediate state of order between gases and crystalline solids (solids can also be disordered; *see* later).

4.2 Gases

Temperature and pressure changes affect the volumes of liquids and solids only slightly, but gas volumes are affected to a much greater degree. A study of these changes led to the gas laws.

4.2.1 BOYLE'S LAW

This states that for a fixed mass of gas at constant temperature, the pressure and volume are inversely proportional, i.e., $V \propto 1/P$ or $PV = \text{constant}$ where P and V represent pressure and volume respectively.

4.2.2 CHARLES–GAY-LUSSAC LAW

It was found that gases expand or contract by approximately 1/273 of their volume at 0 °C for every °C that the temperature is raised or lowered respectively. Consequently, if the temperature of the gas is lowered by 273.15 °C then, theoretically, the volume will be zero; hence, –273.15 °C or 0 K (Kelvin), is known as ABSOLUTE ZERO. 0 °C is, therefore, 273.15 K or, in general, the absolute temperature (K) = 273.15 + t(°C). (Note, 273.15 is usually taken as 273.)

The Charles–Gay-Lussac law, therefore, states that the volume of a fixed mass of gas, at constant pressure, is directly proportional to the absolute temperature, i.e., $V \propto T$ or $V/T =$ constant where $V =$ volume and $T =$ absolute temperature.

4.2.3 IDEAL GAS EQUATION

Boyle's law and the Charles–Gay-Lussac law may be combined, since $V \propto 1/P$ and $V \propto T$, and so $V \propto T/P$, or $V = kT/P$ or $PV/T = k$ (hence the well known formula $P_1 V_1/T_1 = P_2 V_2/T_2$). The constant k depends on the amount of gas; for 1 mole of gas, then $PV/T = R$ where R is called the GAS CONSTANT, and is constant for all gases. For n moles of gas, then $PV = nRT$ and this equation is known as the IDEAL GAS EQUATION, and a gas which obeys this equation is said to be an IDEAL GAS. The value of the gas constant, R, can be calculated. For 1 mole of gas, $n = 1$ and hence $R = PV/T$. Since $P =$ force/area, then

$$R = \frac{\text{force}}{\text{area}} \times \text{volume} \times \frac{1}{T}$$

$$= \text{force} \times \text{length} \times (1/T)$$

But force × length = energy, and the SI unit for energy is the joule (J). Hence the units of R are expressed in joules per mole per degree absolute, i.e., J mol^{-1} K^{-1}.

Substituting the values of P, V and T into the equation $R = PV/T$ will give a value for R:

$P =$ standard pressure, 1 atmosphere, or in SI units is the force exerted by a 0.76 m column of mercury at 0 °C on an area of 1 m^2, and since mercury has a density of 13 600 kg m^{-3}, the column has a mass of $0.76 \times 1 \times 13\,600$ kg. Since the acceleration due to gravity is 9.81 m s^{-2}, the pressure in newtons per square metre is $0.76 \times 13\,600 \times 9.81$ N m^{-2}.
$V =$ volume of 1 mole of gas at standard temperature and pressure (s.t.p.) = 22 400 cm^3 = 0.0224 m^3
$T =$ standard temperature = 0 °C = 273 K

$$R = \frac{PV}{T} = \frac{0.76 \times 13\,600 \times 9.81 \times 0.0224}{273} = 8.32 \text{ J mol}^{-1} \text{ K}^{-1}$$

4.2.4 DALTON'S LAW OF PARTIAL PRESSURES

This states that the total pressure, P, exerted by a mixture of gases is given by the sum of the pressures each gas would exert if it alone occupied the volume of the mixture at the same temperature. Mathematically:

$$P = P_A + P_B + \ldots\ldots\ldots$$

The pressure each constituent gas exerts (e.g., P_A, P_B, etc) is called its PARTIAL PRESSURE and partial pressure

$$P_A = \frac{\text{number moles gas A}}{\text{total number moles gas in mixture}} \times P$$

Numerical Example: An enclosed vessel contains 3.2 g of oxygen and 0.4 g of hydrogen at atmospheric pressure and 0 °C. What will be the partial pressure of oxygen if the temperature is raised to 150 °C? (A_r : O = 16, H = 1). Let total pressure of gas mixture be P_2 at 150 °C. Since volume is constant,

$$P_1/T_1 = P_2/T_2$$

Therefore

$$\frac{760}{273} = \frac{P_2}{423} \quad \text{or} \quad P_2 = \frac{760 \times 423}{273}$$

Hence

$$P_2 = 1177.58 \text{ mmHg}$$

The partial pressure of oxygen

$$p_{O_2} = \frac{\text{moles of } O_2}{\text{moles of } O_2 + \text{moles of } H_2} \times \text{total pressure}$$

$$= \frac{3.2/32}{3.2/32 + 0.4/2} \times 1177.58 = \frac{0.1}{0.3} \times 1177.58$$

$$= 392.53 \text{ mmHg}$$

4.2.5 VAPOUR DENSITY

The vapour density of a gas or vapour is given by

$$\text{Vapour density} = \frac{\text{Mass of 1 volume of gas or vapour}}{\text{Mass of 1 volume of hydrogen at the same temperature and pressure}}$$

Since equal volumes of all gases under the same conditions of temperature and pressure contain the same number of molecules (AVOGADRO'S HYPOTHESIS), then:

$$\text{Vapour density} = \frac{\text{Mass of } n \text{ molecules of gas}}{\text{Mass of } n \text{ molecules of hydrogen}}$$

$$= \frac{\text{Mass of 1 molecule of gas}}{\text{Mass of 1 molecule of hydrogen}}$$

$$= \frac{\text{Mass of 1 molecule of gas}}{\text{Mass of 2 atoms of hydrogen}}$$

Using the old relative mass scale of H = 1, and that

$$\text{Relative molecular mass} = \frac{\text{Mass of 1 molecule of gas}}{\text{Mass of 1 atom of hydrogen}}$$

then

$$\text{Vapour density} = \frac{1}{2} \times \text{Relative molecular mass}$$

or

$$\text{Relative molecular mass} = 2 \times \text{Vapour density}$$

Hence relative molecular masses of gases or vapours can be found from measurements of vapour density.

4.2.6 DETERMINATION OF RELATIVE MOLECULAR MASSES OF GASES AND VAPOURS FROM VOLATILE LIQUIDS

The principle is that of finding the volume of vapour obtained from a known mass of liquid. One form of equipment is shown in *Figure 4.1*. A glass syringe containing a small amount of air is left inside a heating jacket until the volume of air reaches a steady value; this volume is noted. A known mass of the volatile liquid under test is then injected into the glass syringe from a hypodermic syringe.

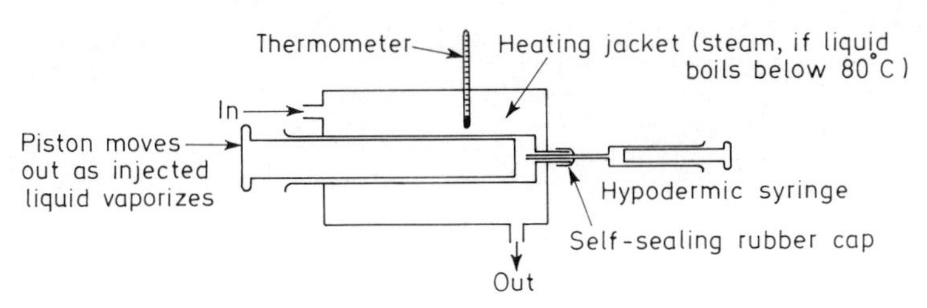

Figure 4.1

The volume of air plus vapour produced is recorded when the reading becomes steady. After correcting the volume of vapour (obtained by subtraction) to s.t.p., the mass of liquid which would yield 22 400 cm^3 of vapour, at s.t.p., is calculated, and hence the relative molecular mass of the test substance is obtained.

4.2.7 KINETIC THEORY OF GASES

In the gaseous state, molecules (or atoms) are in continuous, rapid and random motion (hence diffusion), colliding with themselves and with the walls of the container. From the kinetic theory of gases, the following expression can be derived:

$$PV = \tfrac{1}{3}Nm\bar{c}^2 \tag{4.1}$$

where P = pressure, V = volume, N = number of molecules, m = mass of molecule and $\bar{c}$ = root mean square velocity (i.e., if three molecules have velocities 2, 4 and 6 m s^{-1}, the average velocity is 4 m s^{-1} but the mean square velocity, $\bar{c}^2 = (2^2 + 4^2 + 6^2)/3 = 18.67$, and hence $\bar{c} = \sqrt{18.67} = 4.32$ m s^{-1}).

Equation (4.1) holds for ideal gases, and in deriving this equation, a number of assumptions are made:

(1) Collisions between molecules and between molecules and walls of containing vessel are perfectly elastic (*see* below).
(2) Forces between molecules or between molecules and walls of containing vessel are negligible.
(3) The volume of the molecules is negligible compared with the volume of the containing vessel.

From equation (4.1) it is possible to derive all the gas laws, including the ideal gas equation, $PV = nRT$, and Graham's law of diffusion.

4.2.7.1 Graham's Law of Diffusion

This law states that the rate of diffusion of a gas is inversely proportional to the square root of its density. The proof is as follows:

From equation (4.1)

$$\bar{c} = \sqrt{\frac{3PV}{Nm}}$$

But density

$$d = \frac{\text{Mass}}{\text{Volume}} = \frac{Nm}{V}$$

so

$$\bar{c} = \sqrt{3P/d}$$

Now the more rapidly the molecules move, the more rapidly they diffuse, i.e.,

$$\bar{c} \propto \text{Rate of diffusion}$$

Therefore

Rate of diffusion $= \sqrt{k/d}$ at the same temperature and pressure, where k is a constant.

Hence

$$\text{Rate of diffusion} \propto \sqrt{1/d}$$

For two gases, A and B, at the same temperature and pressure, the equation becomes

$$\frac{\text{Rate of diffusion of A}}{\text{Rate of diffusion of B}} = \sqrt{\frac{d_B}{d_A}}$$

Because relative molecular mass is proportional to density (*see* before), it follows that a heavy gas will diffuse more slowly than a light gas. Diffusion

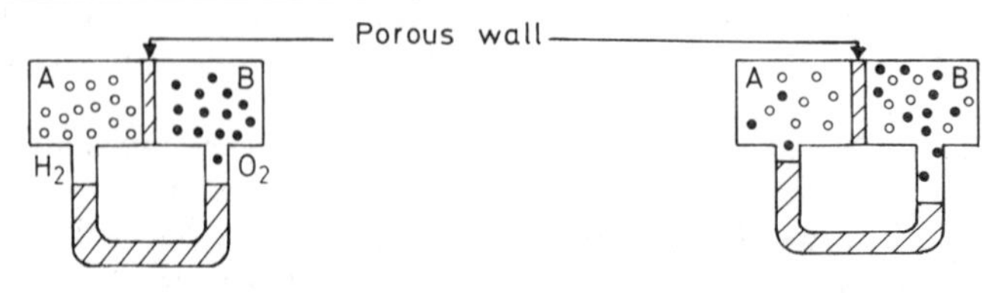

(a) Initial state (b) Short time later

Figure 4.2

occurs because of the kinetic energy of motion through space (TRANSLATIONAL ENERGY) of the gas molecules, and can be demonstrated by the apparatus shown in *Figure 4.2.* This shows hydrogen and oxygen, initially at the same temperature and pressure, separated by a porous wall which allows passage of molecules. After a short time, more hydrogen molecules will have diffused from A to B than oxygen molecules from B to A – hence the pressure in B increases.

4.2.8 DEVIATIONS FROM IDEAL BEHAVIOUR (I.E., REAL OR NON-IDEAL GASES)

In the kinetic theory of gases, several simplifying assumptions were made. Hence the gas behaved as an ideal gas, and so the gas laws and ideal gas equation refer to ideal gases.

Assumptions (2) and (3) are more likely to hold when the concentration of gas is low, i.e., low pressure, since the average volume available for each molecule is large. As pressure is increased, or as temperature is decreased, deviation from the ideal gas situation occurs. Molecules now have more attraction for each other, and their volume is no longer negligible compared with the volume of the container, i.e., liquefaction is approached. The attractive forces operating can include van der Waals, dipole–dipole or hydrogen bonding (*see* Chapter 3). In assumption (1), collisions are perfectly elastic which means that, in a collision, the molecule with the higher translational energy transfers part of this energy to the molecule of lower energy, the latter increasing its own translational energy by just this amount. If, however, the collision is inelastic, the molecule receiving the energy can use it by increasing its own INTERNAL energy instead, i.e., for VIBRATIONS and ROTATIONS, and the molecule may bend or distort. Clearly, the number and rigidity of bonds within a molecule will influence how much the gas deviates from elastic collisions – a molecule with a large number of atoms and therefore bonds, within itself, will in general be more deformable and therefore less elastic in a collision. Hence, monatomic gases (e.g., noble gases) tend to obey the ideal gas equation more closely than, say, polyatomic gas molecules (also, noble gases have only weak forces of attraction, even at high pressure; *see* Chapter 3). However, despite the above arguments there is a limited range of P, V and T where the ideal gas equation holds reasonably well (for any particular gas).

4.2.9 MOLECULAR VELOCITIES

At a particular temperature, the molecules of a gas have a wide range of velocities because of frequent collisions (*Figure 4.3*). Relatively few molecules have very low or very high velocities, most having intermediate velocities. An increase in temperature gives a general shift to higher velocities with a significant increase in the number of molecules having very high velocities.

The range of molecular velocities can be demonstrated and determined by the Zartmann experiment. A narrow beam of tin molecules is directed through a vacuum at a rotating disc, which is split into numbered segments. Every time the disc comes round to the first segment, another beam of tin molecules is directed at the rotating disc. Eventually, a layer of tin builds up on the disc, fast moving molecules having landed on low numbered segments, and slow moving molecules

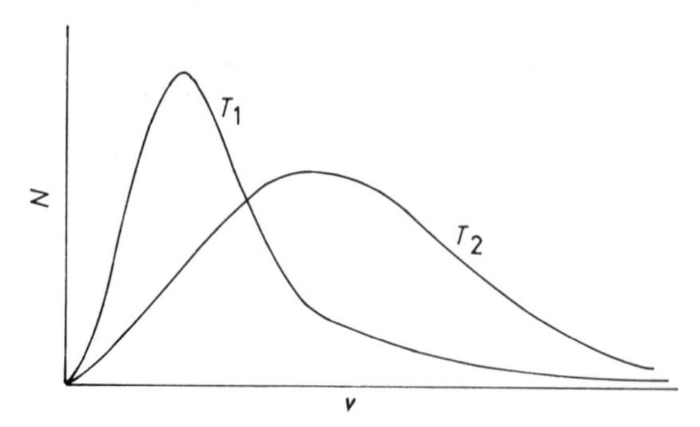

Figure 4.3 A plot of N, the number of molecules having a particular velocity, against v, the molecular velocity, at two temperatures, where T_2 is higher than T_1

on high numbered segments. The masses of tin on the various segments correspond to the spread of molecular velocities.

4.3 Liquids

By decreasing the temperature or increasing the pressure, gases may be liquefied. However, for every gas there is a temperature, called the CRITICAL TEMPERATURE, above which application of pressure alone, no matter how large, fails to liquefy the gas. The pressure needed to cause liquefaction at the critical temperature is called the CRITICAL PRESSURE.

Molecules in a liquid are continually moving and colliding. This can be demonstrated by the phenomenon known as BROWNIAN MOVEMENT. This is the random movement of small suspended particles (e.g., pollen grains in water), observable under a microscope, caused by their being constantly bombarded by solvent molecules. In the liquid, some molecules will have higher than average velocities, and if they are near the surface they may overcome attractive forces to other liquid molecules and hence leave the liquid to enter the gas phase, i.e., evaporation occurs. If no heat is supplied the liquid's temperature falls. The amount of heat needed to change one mole of liquid to vapour without change in temperature is the LATENT HEAT OF VAPORIZATION.

A property of liquid surfaces is that they tend to contract to the smallest possible area, e.g., drops of liquid become spherical. In the interior of a liquid, any molecule is surrounded by other molecules and so, on average, is attracted equally in all directions. A molecule on the surface, however, experiences a resultant attraction inwards; consequently, the surface behaves as if it were in a state of tension. This surface property is called SURFACE TENSION.

4.4 Solids

In the solid state, molecules, atoms or ions do not move randomly throughout the material. However, they are not stationary since the solid does have kinetic energy because of LATTICE VIBRATIONS. The term LATTICE refers to an ordered arrangement of molecules (or atoms or ions) and in this state materials are said to be CRYSTALLINE.

When the temperature of the solid is raised, the molecules vibrate even more about their mean positions in the lattice. Eventually, the lattice vibrations may reach such a level that molecules may move (or translate) away from their lattice sites. This weakens forces on other neighbouring molecules, and hence the latter are even easier to displace from their lattice sites. This is MELTING, and the heat required to melt one mole of the substance without change in temperature is the LATENT HEAT OF FUSION. Some solids, when heated, turn directly to vapour, thus leaving out the liquid state (because the liquid's vapour pressure at the melting point is greater than atmospheric pressure – *see* also Chapter 6). This is known as SUBLIMATION.

4.4.1 EXPERIMENTAL METHOD FOR DETERMINING STRUCTURES

X-*Ray Diffraction*

The arrangement of particles in a solid can be determined by X-ray diffraction. This is because the wavelength of X-rays and interatomic distances in crystals are of similar magnitude ($\sim 10^{-10}$ m) and so the regular arrangement of particles in the crystal can act as a diffraction grating to X-rays.

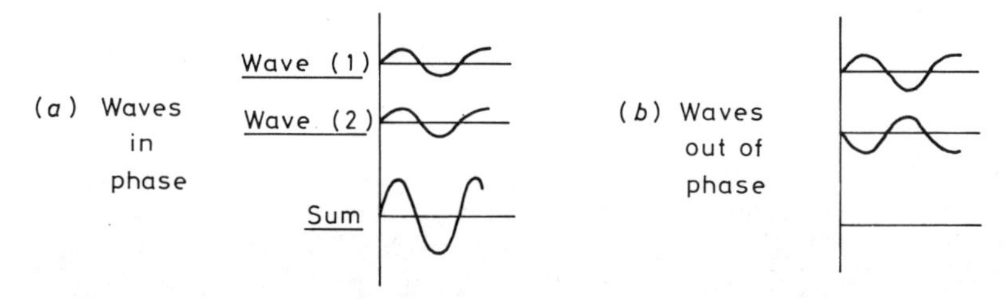

Figure 4.4 Constructive (a) and destructive (b) interference in wave motion

X-Rays can CONSTRUCTIVELY or DESTRUCTIVELY interfere with one another [*Figure 4.4(a)* and *4.4(b)* respectively]. In the case of constructive interference, the beam is reinforced. If a crystal lattice, in which identical atoms lie in parallel planes 1, 2, 3, 4 etc., is subjected to a beam of X-rays of wavelength λ, then some of the radiation will be reflected from the planes (*Figure 4.5*). If the reflected waves are still to be in phase, the extra distance travelled by the lower wave must be equal to a whole number of wavelengths, i.e.,

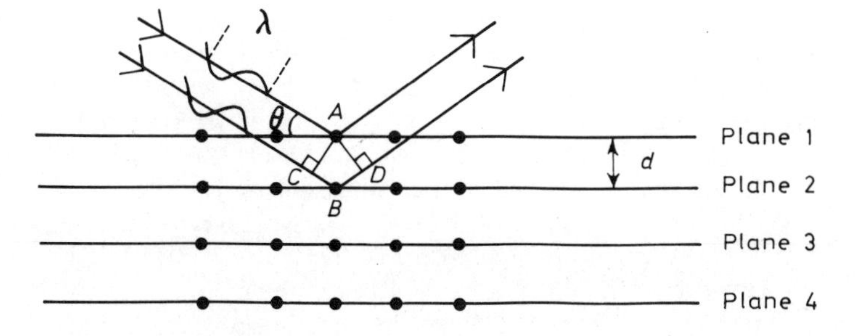

Figure 4.5 Reflection of X-*rays by crystal planes*

$CB + BD = n\lambda$, where n = an integer. Since $AB = d$, then $CB = BD = d \sin\theta$, or $CB + BD = 2d \sin\theta$.

Hence:

$$2d \sin\theta = n\lambda \qquad (4.2)$$

Equation (4.2) is called the Bragg equation. Maximum reinforcement occurs for certain values of θ (corresponding to n = 1, 2, 3 etc.,) and if these angles are found experimentally and λ is known, then the lattice spacing d can be calculated using equation (4.2). The diffracted *X*-rays are detected by their diffraction pattern on a photographic plate, which is a pattern of spots of high intensity when a crystal is used.

In any crystal, the occupants of the lattice sites may be individual atoms, ions or molecules. Since it is the electrons which are responsible for the diffraction or scattering of *X*-rays, the intensity of the diffraction pattern depends on the number and distribution of electrons at lattice sites. The structure of molecules at lattice sites can therefore be determined.

Electron diffraction can be used in a similar manner. A beam of electrons is passed through a gas, and the beam is diffracted by the atoms in the molecules of the gas. The diffraction pattern produced gives information concerning the structure of the molecules.

4.4.2 METALLIC BONDING

In metals, whilst the atoms occupy lattice sites, the valency electrons are not localized in orbitals on any particular atom, but are delocalized into orbitals (or BANDS) which extend over the whole of the material. This leads to a lowering in energy for the system (rather like delocalization stabilizes benzene – *see* Chapter 2, p. 22 – but on a much larger scale), and it is this which results in the metal atoms being bonded together. The nature of these delocalized, or freely moving, electrons explains the high electrical and thermal conductivity of metals, their metallic lustre and the relative ease with which electrons are removed from their surfaces by light or heat. Because of the mobile valency electrons, bonds are not strongly directional and so one plane of metal atoms may slip over another; hence metals are ductile and malleable.

4.4.3 STRUCTURES OF METAL CRYSTALS

Metal crystals can generally be thought of as a three dimensional structure formed by the closest packing of spheres [*see Figure 4.6(a)*]. If a layer of spheres is close packed, and a second close packed layer is then placed on top of it [*Figure 4.6(b)*], then two types of hollow in the second layer exist. If the third layer is placed in the hollows marked *H*, then these spheres lie directly above the spheres of the first layer. The fourth layer is formed by repeating the second layer and so a repeating pattern *ABAB* results; this is called HEXAGONAL CLOSE PACKING. However, if the third layer is placed in the hollows marked *C*, a new layer is formed in that its spheres do not lie directly above any of the spheres in the first two layers, and so a repeating pattern *ABCABC* is obtained.

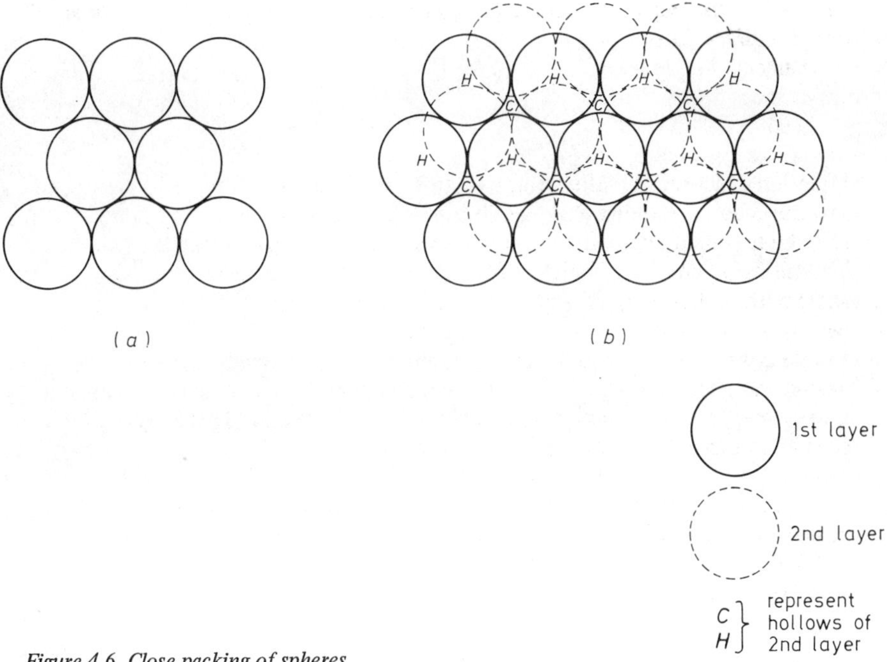

Figure 4.6 Close packing of spheres

This is called CUBIC CLOSE PACKING, and a closer study of this type of packing shows that the crystal is made up of many identical UNIT CELLS (repeating units); one of these is shown in *Figure 4.7(a)* and is called the FACE-CENTRED CUBE.

Figure 4.7 (a) Cubic close packing unit cell (face-centred cube); (b) Body centred cubic packing unit cell

There are a few properties which are common to both of the closest packed structures (cubic and hexagonal).

(1) In both cases, 74 per cent of the available space is occupied by spheres.
(2) In either of the closest packed structures each sphere is in contact with 12 nearest neighbours: six lie in one layer and three from each of the layers above and below [*Figure 4.6(b)*]. The number of nearest neighbours is called the CO-ORDINATION NUMBER of the sphere.

Several metals are found in the BODY-CENTRED cubic structure [*Figure 4.7(b)*], but since each sphere has a co-ordination number of only 8, it is not a close packed structure, i.e., this is a less efficient form of packing, the spheres occupying only 68 per cent of the available space.

Nearly all metals crystallize into one or more of the above structures. Examples include:

(1) Hexagonal close packing: Mg, Zn.
(2) Cubic close packing: Al, Cu, Ag, Au.
(3) Body centred cubic packing: Group I alkali metals.

4.4.4 IONIC LATTICES

In ionic crystals, the repeating units of the lattice are positively and negatively charged ions. The high melting points characteristic of ionic crystals are because of the powerful electrostatic forces holding the ions together in the crystal. Some ionic lattice structures are now discussed.

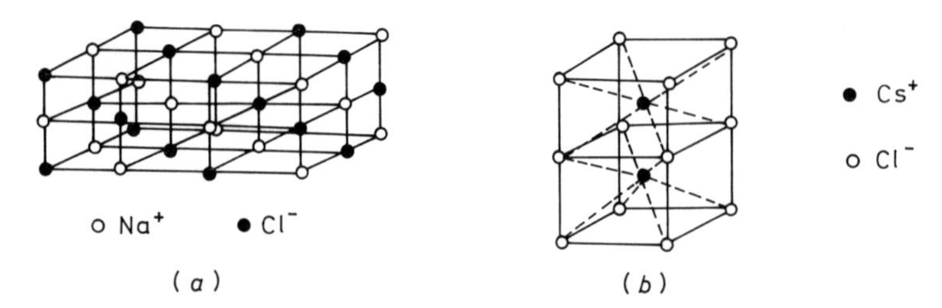

Figure 4.8 Sodium chloride and cæsium chloride crystal lattices

The sodium chloride lattice is shown in *Figure 4.8(a)*, which is made up of a face-centred cube of chloride ions and a face-centred cube of sodium ions, which are interpenetrating. Moreover, each Na^+ ion has six nearest neighbour Cl^- ions, and each Cl^- ion has six nearest neighbour Na^+ ions; hence, the co-ordination is referred to as 6:6. For caesium chloride [*Figure 4.8(b)*], both the Cs^+ and Cl^- ions have co-ordination numbers of 8, and so the co-ordination is 8:8. If *Figure 4.8(b)* is extended, it will be seen that the lattice is made up of a simple cube of Cl^- ions, and a simple cube of Cs^+ ions, which are interpenetrating. In the zinc sulphide lattice, each Zn^{2+} ion is surrounded tetrahedrally by four S^{2-} ions, and *vice versa*, giving 4:4 co-ordination. Finally, for the calcium fluoride lattice (fluorite), each Ca^{2+} ion is surrounded by eight F^- ions at the corners of a cube, and each F^- ion is surrounded by four Ca^{2+} ions at the corners of a tetrahedron. This is an example of 8:4 co-ordination.

4.4.5 COVALENT NETWORK SOLIDS

Crystals in which all the atoms are linked by a continuous system of covalent bonds are called covalent network solids. For example, diamond is made of carbon atoms (sp^3 hybridized) covalently bonded, in a tetrahedral configuration, to all the other carbon atoms in the lattice [*see Figure 4.9(a)*]. Hence diamond is one of the hardest substances known (it is also an electrical insulator).

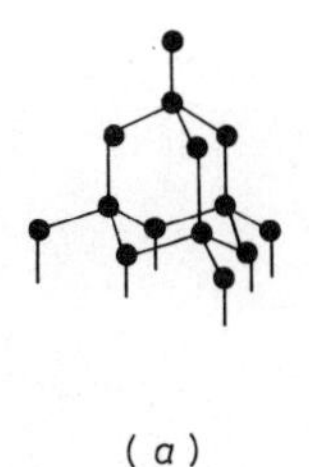

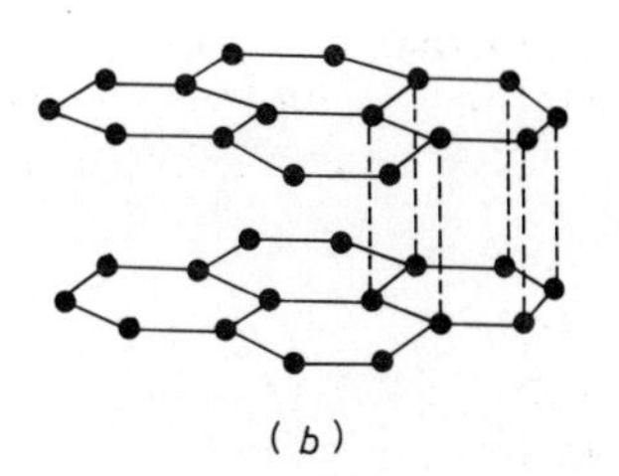

Figure 4.9 (a) Three-dimensional lattice in diamond. (b) Two-dimensional lattice in graphite

Graphite consists of sheets of carbon atoms [*Figure 4.9(b)*]. Each carbon atom is sp^2 hybridized and so a planar network of hexagons results. The unpaired $2p$ electron on each carbon atom (left after hybridization) can now move through the sheets *via* adjacent overlapping $2p$ orbitals; electrons are therefore delocalized over the layer. The electrical conduction within layers is therefore high, but across the layers (i.e., perpendicular to the layers) it is low, since the layers are bonded only by weak van der Waals forces. This also explains the smooth feel of graphite, since these weak forces allow the sheets or layers of carbon atoms to slide over one another.

4.4.6 OTHER FORCES IN SOLIDS

Other forces operating in solids include van der Waals forces, dipole–dipole forces and hydrogen bonding (*see* also Chapter 3).

Van der Waals forces play an important part in molecular crystals such as iodine or naphthalene. It is these weak forces which hold the lattice together. Van der Waals forces are the only forces which operate between atoms in the solid state of the noble gases (giving close packed structures), and this is why the noble gases have such low freezing points (e.g., He has the lowest freezing point of any material at 1.75 K).

Hydrogen bonding plays an important part in the lattice structure of salt hydrates, e.g., gypsum, $CaSO_4.2H_2O$, in which layers of calcium sulphate are held together by hydrogen bonds with the water molecules. Therefore, the crystals cleave readily along these layers because the hydrogen bonds are much weaker than the ionic attractions. Anhydrite, $CaSO_4$, does not suffer from this, however, since no hydrogen bonds are present. Other examples of hydrogen bonding in solids include ice and proteins (*see* Chapter 3).

4.4.7 NON-CRYSTALLINE SOLIDS

The molecular arrangement in solids does not have to be regular as in a lattice; if the arrangement is irregular, the solid is said to be AMORPHOUS and no regular *X*-ray diffraction pattern is observed. This situation occurs particularly for large irregular molecules such as polymers or MACROMOLECULES. Here, the chains can line up regularly (crystalline) or irregularly (amorphous) and this is shown in *Figure 4.10* for poly(ethene).

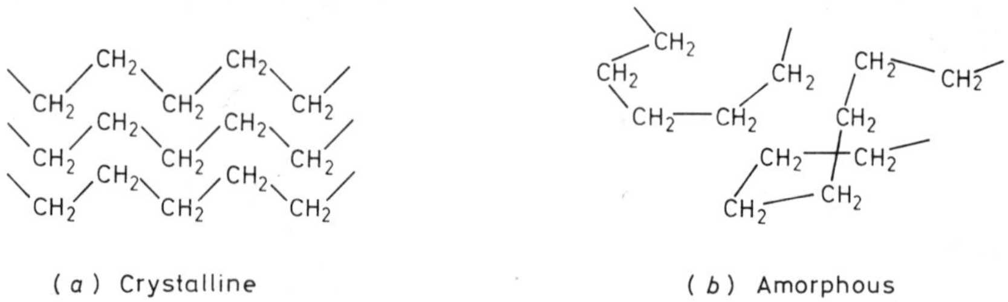

(*a*) Crystalline

(*b*) Amorphous

Figure 4.10 Crystalline and amorphous packing of chains

5

ENERGETICS

In virtually all chemical reactions, there are energy changes; the study of these energy changes is called ENERGETICS.

5.1 The First Law of Thermodynamics

This states that although energy may be converted from one form to another, it cannot be created or destroyed (which is equivalent to the law of conservation of energy). The first law is expressed mathematically as

$$\Delta U = q - w \tag{5.1}$$

where ΔU is the change in internal energy of the system, q is the total heat absorbed by the system and w is the work done by the system. (N.B. If heat is evolved then q is negative, and if work is done on the system then w is negative.)

In most chemical reactions, the pressure remains constant, i.e., reaction may be carried out in an open beaker. Work can be done by a gas or other system expanding against the atmosphere (e.g., zinc + dilute hydrochloric acid, where the hydrogen evolved has to push back the atmosphere). This work term, w, can be calculated by considering a cylindrical vessel fitted with a piston (to allow expansion), containing some material which may be a gas, liquid or solid (or a mixture of these phases). If the material expands, pushing back the piston, then:

$$\begin{aligned} w &= \text{force on piston} \times \text{distance moved} \\ &= P \times A \times \text{distance moved} = P\Delta V \end{aligned}$$

where P = external pressure on piston, A = surface area of piston, and ΔV = change in volume. Substituting into equation (5.1) gives:

$$\Delta U = q - P\Delta V \quad \text{or} \quad q = \Delta U + P\Delta V \tag{5.2}$$

(provided only work due to expansion occurs).

At CONSTANT VOLUME, $\Delta V = 0$, and hence $\Delta U = q_v$, i.e., the heat absorbed at constant volume, q_v, equals the change in internal energy of the system.

At CONSTANT PRESSURE, from equation (5.2), $q_p = \Delta U + P\Delta V$, i.e., the heat absorbed by the system at constant pressure, q_p, is equal to the change in ENTHALPY or heat content, ΔH, where $\Delta H = \Delta U + P\Delta V = q_p$.

Internal energy change is less useful than enthalpy change since reactions are not usually carried out at constant volume. For reactions in which liquids and solids are involved, very little volume change occurs, and so if the reaction occurs at atmospheric pressure the $P\Delta V$ term is small and so $\Delta H \approx \Delta U$. For gas reactions, however, ΔH and ΔU can differ considerably. The enthalpy change (or heat of reaction), ΔH, is given by

$$\Delta H = \begin{pmatrix}\text{Enthalpy of products at} \\ \text{temperature } T \text{ and} \\ \text{atmospheric pressure}\end{pmatrix} - \begin{pmatrix}\text{Enthalpy of reactants at} \\ \text{temperature } T \text{ and} \\ \text{atmosphere pressure}\end{pmatrix}$$

Consequently, if a reaction is exothermic, then enthalpy of reactants, H_1, must be greater than that of products, H_2, so the enthalpy change ΔH ($=H_2 - H_1$) is therefore negative. It therefore follows that for an endothermic reaction, ΔH is positive.

The enthalpy change is usually determined by insulating the system from its surroundings and allowing the heat of reaction to alter the system's temperature. The amount of heat which has to be put in or taken out of the system to restore it to its original temperature (i.e., ΔH) is then calculated.

5.1.1 STANDARD ENTHALPY CHANGES

In the work to follow, the subscripts (g), (s), (l) and (aq) will denote gas, solid, liquid and aqueous states respectively.

For the reaction:

$$\text{C (graphite)} + O_{2\,(g)} \rightarrow CO_{2\,(g)}; \ \Delta H = -393.5 \text{ kJ mol}^{-1}$$

The enthalpy change refers to the amounts shown in the equation, i.e., one mole each of carbon, oxygen and carbon dioxide. Normally, STANDARD ENTHALPY changes are quoted, which refer to standard pressure and some temperature, generally 25 °C (298 K); these are written as $\Delta H^{\ominus}_{298}$. This symbol also means that the substances must be in their normal physical states under these conditions, so the above example refers to the most stable allotrope of carbon (graphite) and gaseous oxygen and carbon dioxide. If solutions are involved they must be at unit activity (which approximates to 1 mole per litre). If a reaction does not occur under standard conditions, the enthalpy change is determined at a temperature at which reaction does occur, and can then be corrected to standard conditions.

The enthalpy change when one mole of a substance is formed from its elements in the standard state is known as the STANDARD ENTHALPY (HEAT) OF FORMATION, and is given the symbol $\Delta H^{\ominus}_{f,298}$. In the above example, the enthalpy change ΔH, is therefore the standard enthalpy (heat) of formation of carbon dioxide.

The STANDARD ENTHALPY (HEAT) OF COMBUSTION of a substance $\Delta H^{\ominus}_{c,298}$, refers to the enthalpy change when one mole of it undergoes complete combustion in the standard state (e.g., in a compound containing carbon, the carbon must be oxidized to carbon dioxide, not carbon monoxide).

Example:

$$CH_{4\,(g)} + 2O_{2\,(g)} \rightarrow CO_{2\,(g)} + 2H_2O_{(l)}; \ \Delta H^{\ominus}_{c,298} = -890.4 \text{ kJ mol}^{-1}$$

The STANDARD ENTHALPY (HEAT) OF ATOMIZATION, $\Delta H^{\ominus}_{at,298}$ of an element refers to the enthalpy change when one mole of gaseous atoms is formed from the element in the standard state.

Example:

$\frac{1}{2}O_{2\,(g)} \rightarrow O_{(g)};\ \Delta H^{\ominus}_{at,298} = +249.2\ \text{kJ mol}^{-1}$

The STANDARD ENTHALPY (HEAT) OF HYDROGENATION refers to the enthalpy change involved when one mole of an unsaturated material is completely hydrogenated in the standard state.

Example:

$C_2H_{2\,(g)} + 2H_{2\,(g)} \rightarrow C_2H_{6\,(g)};\ \Delta H^{\ominus}_{298} = -311.5\ \text{kJ mol}^{-1}$

An interesting case is the enthalpy of hydrogenation of benzene:

$C_6H_{6\,(l)} + 3H_{2\,(g)} \rightarrow C_6H_{12\,(l)};\ \Delta H^{\ominus}_{298} = -209\ \text{kJ mol}^{-1}$

whereas for cyclohexene:

$C_6H_{10\,(l)} + H_{2\,(g)} \rightarrow C_6H_{12\,(l)};\ \Delta H^{\ominus}_{298} = -120\ \text{kJ mol}^{-1}$

Benzene might be expected, therefore, to have an enthalpy of hydrogenation of $3 \times -120 = -360\ \text{kJ mol}^{-1}$ which, compared with the experimental value, gives a difference of $151\ \text{kJ mol}^{-1}$. This difference indicates the stability of the benzene structure, and can be explained by resonance theory (*see* Chapter 2). This is an example of how enthalpy changes can provide information concerning structure or bonding in some compounds.

5.1.2 HESS'S LAW

This states that the total energy change in a chemical reaction is independent of the route by which the chemical reaction takes place. It is very useful for calculating enthalpy changes, for reactions, which cannot be determined by direct measurement.

Example: Calculate the heat of formation of ethane from the following heats of combustion: Carbon $= -393.5\ \text{kJ mol}^{-1}$; hydrogen $= -286\ \text{kJ mol}^{-1}$; ethane $= -1560\ \text{kJ mol}^{-1}$

These quantities can be represented as follows:

(1) $C + O_2 \rightarrow CO_2;\ \Delta H^{\ominus}_{c,298} = -393.5\ \text{kJ mol}^{-1}$

(2) $H_2 + \frac{1}{2}O_2 \rightarrow H_2O;\ \Delta H^{\ominus}_{c,298} = -286\ \text{kJ mol}^{-1}$

(3) $C_2H_6 + 3\frac{1}{2}O_2 \rightarrow 2CO_2 + 3H_2O;\ \Delta H^{\ominus}_{c,298} = -1560\ \text{kJ mol}^{-1}$

If (1) is multiplied by 2, and (2) multiplied by 3 is added, and then (3) is subtracted, it is seen that:

$$
\begin{array}{lcl}
2C + 2O_2 & \rightarrow & 2CO_2 \\
3H_2 + 1\tfrac{1}{2}O_2 & \rightarrow & 3H_2O \\
-C_2H_6 - 3\tfrac{1}{2}O_2 & \rightarrow & -2CO_2 - 3H_2O \\
\hline
2C + 3H_2 & \rightarrow & C_2H_6
\end{array}
$$

Hence

$$\Delta H^{\ominus}_{f,298}(\text{ethane}) = (2 \times -393.5) + (3 \times -286) + 1560 = -85 \text{ kJ mol}^{-1}$$

5.1.3 BOND ENERGIES

Consider the process:

$$CH_{4\,(g)} \rightarrow C_{(g)} + 4H_{(g)}$$

The enthalpy change for this process, $\Delta H^{\ominus}_{298}(1)$, can be obtained from the following data:

(1) $C_{(graphite)} + 2H_{2\,(g)} \rightarrow CH_{4\,(g)};\ \Delta H^{\ominus}_{f,298} = -75 \text{ kJ mol}^{-1}$

(2) $2H_{2\,(g)} \rightarrow 4H_{(g)};\ \Delta H^{\ominus}_{298} = 4 \times \Delta H^{\ominus}_{at,298} = 4 \times 218 = 872 \text{ kJ mol}^{-1}$

(3) $C_{(graphite)} \rightarrow C_{(g)};\ \Delta H^{\ominus}_{at,298} = 715 \text{ kJ mol}^{-1}$

$$\Delta H^{\ominus}_{298}(1) = -(1) + (2) + (3) = 75 + 872 + 715 = 1662 \text{ kJ mol}^{-1}$$

Each C–H bond would therefore be expected to make an energy contribution of $1662/4 = 415.5 \text{ kJ mol}^{-1}$ to the total enthalpy change, since the four bonds are equivalent. This value for the C–H bond varies from compound to compound, i.e., C–H bonds do not have exactly the same energy associated with them in, say, ethane and benzene. However, average standard bond enthalpies or AVERAGE BOND ENERGIES for particular types of bond, are very useful quantities; they can be used, for example, in estimating enthalpy changes in reactions.

Example: Calculate the approximate enthalpy change, ΔH, involved when ethane is dissociated into its atoms, given that the average bond energies of C–C and C–H bonds are 346 and 414 kJ mol^{-1} respectively.

Hence, enthalpy change, $\Delta H = 346 + (6 \times 414) = 2830 \text{ kJ mol}^{-1}$

5.1.4 THE BORN–HABER CYCLE

The LATTICE ENERGY of an ionic crystal can be defined as the standard enthalpy (heat) of formation of one mole of the crystal lattice from its constituent ions in the gas phase.

$$K^+_{(g)} + Cl^-_{(g)} \rightarrow K^+Cl^-_{(s)};\ \Delta H^{\ominus}_{298} = \text{lattice energy}$$

Lattice energies cannot be determined directly, but can be found by applying Hess's law to an energy cycle known as a Born–Haber cycle (*Figure 5.1*).

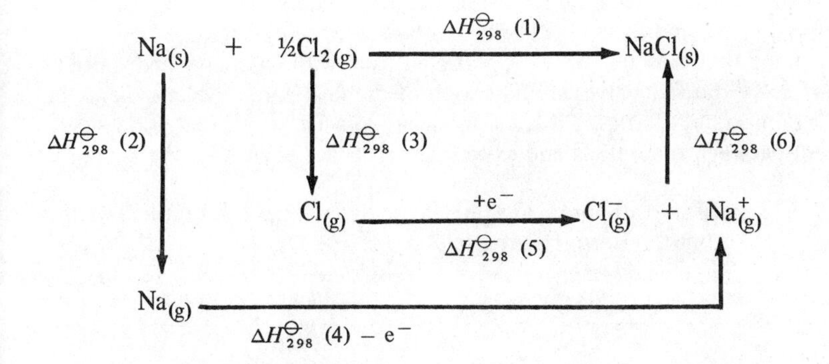

Figure 5.1 The Born–Haber cycle for sodium chloride

The enthalpy changes are as follows: $\Delta H^\ominus_{298}(1)$ = standard enthalpy (heat) of formation of sodium chloride = –411 kJ mol^{-1}; $\Delta H^\ominus_{298}(2)$ = standard enthalpy (heat) of atomization of sodium = 108.4 kJ mol^{-1}; $\Delta H^\ominus_{298}(3)$ = standard enthalpy (heat) of atomization of chlorine = 121.1 kJ mol^{-1}; $\Delta H^\ominus_{298}(4)$ = ionization energy of sodium = 500 kJ mol^{-1}; $\Delta H^\ominus_{298}(5)$ = electron affinity of chlorine = –364 kJ mol^{-1}; $\Delta H^\ominus_{298}(6)$ = lattice energy (often given the symbol U) of sodium chloride.

By Hess's law:

$$\Delta H^\ominus_{298}(1) = \Delta H^\ominus_{298}(2) + \Delta H^\ominus_{298}(3) + \Delta H^\ominus_{298}(4) + \Delta H^\ominus_{298}(5) + \Delta H^\ominus_{298}(6)$$

or

$$-411 = 108.4 + 121.1 + 500 - 364 + \Delta H^\ominus_{298}(6)$$

Hence,

$$\Delta H^\ominus_{298}(6) = -411 - 108.4 - 121.1 - 500 + 364$$

$$= -1140.5 + 364 = -776.5 \text{ kJ mol}^{-1}$$

i.e., lattice energy of sodium chloride is –776.5 kJ mol^{-1}

A knowledge of lattice energy is useful as the following examples illustrate.

(1) The dependence of melting points of ionic solids on lattice energy is shown in *Table 5.1*, i.e., as expected, the greater the lattice energy, the higher the melting point.

Table 5.1 Lattice energies and melting points

Compound	*Lattice energy*/kJ mol^{-1}	*Melting point*/°C
NaF	–915	995
NaCl	–781	808
NaBr	–743	750
NaI	–699	662

(2) A comparison of experimental values of lattice energy (i.e., from Born–Haber cycles) and theoretical values of lattice energy (calculated by assuming that ions are spherical particles, each with its charge distributed uniformly round it) gives an indication of the degree of covalent character in a compound (*Table 5.2*). It can be seen from *Table 5.2* that good agreement between theoretical and experimental values occurs for the alkali

Table 5.2 Comparison of theoretical values of lattice energy U/kJ mol^{-1} with those calculated from the Born–Haber cycle

Compound	*Theory*	*Born–Haber cycle*
NaCl	−766	−781
NaBr	−731	−743
NaI	−686	−699
KI	−631	−643
AgCl	−769	−890
AgBr	−759	−877
AgI	−736	−867
ZnS	−3427	−3565

metal halides, indicating that the simple model of an ionic crystal is a good one. Substantial discrepancies, however, are seen for the silver halides and zinc sulphide, indicating the presence of some covalent bonding.

(3) Lattice energies can be used, in conjunction with Born–Haber cycles, to explain why some compounds do not exist.

(4) Lattice energies can be a guide to solubility. The energy released when a mole of a substance in the form of gaseous ions is solvated (*see* Chapter 3) is known as the SOLVATION ENERGY, or HYDRATION ENERGY if the solvent is water. The process of dissolving can be represented by an energy cycle (*see Figure 5.2*); in this energy cycle, the ENTHALPY (HEAT)

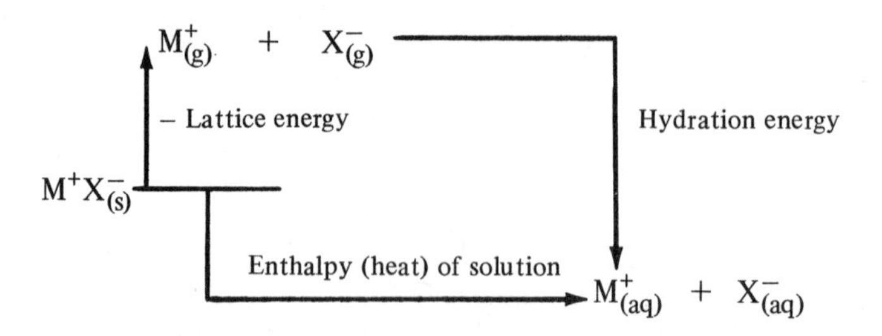

Figure 5.2 Energy cycle for dissolution of an ionic salt in water

OF SOLUTION is the enthalpy (heat) change on dissolving one mole of solute in such an excess of solvent that no further heat change occurs on the addition of more solvent. Solubility occurs when the hydration energy is greater than the lattice energy, resulting in the enthalpy (heat) of solution being negative. However, positive enthalpies of solution also exist, since entropy factors have to be considered (*see* below).

5.2 The Second Law of Thermodynamics: Entropy

One driving force for chemical reaction to occur is for products to have a lower energy state than reactants. However, not all chemical reactions are exothermic; endothermic reactions do occur. Similarly, not all water-soluble substances dissolve in water exothermically, e.g., ammonium chloride dissolves endothermically. Why, then, do some substances, of their own accord (or SPONTANEOUSLY), enter a higher energy state? To explain this, the concept of ENTROPY has to be considered, which can be described as a measure of disorder in a system. Consider two flasks, A and B, connected by a tap, where flask A contains a gas at s.t.p., whilst flask B is evacuated. If the tap is opened, gas will diffuse from A to B until the pressures are the same. The probability of finding all the gas molecules in either flask A or flask B at any time is now negligibly small. Similarly, if some red balls and green balls are shaken together, the balls mix, and unmixing does not occur, even for only a small number of balls. Given a free choice, a system tends to move to a state of more disorder, not less. Consequently, the dissolving of ammonium chloride by water, whilst being unfavourable from energy considerations, does nevertheless occur since a greater state of disorder (or increase in ENTROPY) is attained as the ordered crystal lattice of the salt breaks down. Similarly, many endothermic reactions occur when a gas is one of the products, since the gas state is more disordered than the solid (or liquid) state.

The SECOND LAW OF THERMODYNAMICS states that any system, of its own accord, will always undergo change in such a way as to increase the entropy. The entropy change, ΔS, can be defined as the reversible heat involved in a particular process divided by the temperature (in degrees Kelvin) at which the process occurs, i.e., $\Delta S = q_{rev}/T$. If all the heat supplied to a system is used to create disorder, then none of the heat must be used in raising the system's temperature (if T remains constant, the process is said to be ISOTHERMAL). This situation occurs when a solid melts or a liquid evaporates, and at the melting or boiling point, the addition or subtraction of a small amount of heat energy will reverse the direction of the phase change. These changes are said to be REVERSIBLE changes, the heat involved being q_{rev}.

The general condition for a reaction to be spontaneous is that

$$\Delta S_{total} > 0 \tag{5.3}$$

where $\Delta S_{total} = \Delta S_{system} + \Delta S_{surroundings}$

If the reaction is exothermic, then the heat evolved ($-\Delta H$) is given to the surroundings, e.g., the room, and if the surroundings are sufficiently large so that negligible temperature change occurs, then from equation (5.3):

$$\Delta S_{system} - \Delta H/T > 0$$

or, multiplying through by T we obtain

$$T\Delta S - \Delta H > 0$$

or

$$\Delta H < T\Delta S$$

Therefore, if an exothermic reaction is accompanied by an increase in entropy, the reaction will be spontaneous (since ΔH is negative and $T\Delta S$ is positive); it will also be spontaneous if the entropy decreases, provided that the $T\Delta S$ term is not more negative than ΔH. Endothermic reactions will be spontaneous if the increase in entropy is sufficient to make $T\Delta S$ greater than ΔH, which can often be so at high temperatures.

6

PHASE EQUILIBRIA

6.1 One Component Systems

If a liquid of relatively low boiling point is placed in a container open to the atmosphere, it will eventually evaporate entirely (*see* Chapter 4: liquids). If, however, the liquid is in a closed container, vapour molecules can accumulate in the space above the liquid. As the vapour pressure builds up, gas molecules collide with, and re-enter, the liquid phase, i.e., CONDENSATION occurs. If the system is left at constant temperature, the rate of condensation will become equal to the rate of evaporation. The system is now at equilibrium and the pressure of the vapour remains constant. Since evaporation and condensation do not stop, but proceed at equal rates, the phase equilibrium is described as being DYNAMIC in nature. Experimental data show that the equilibrium vapour pressure of a liquid increases as temperature increases.

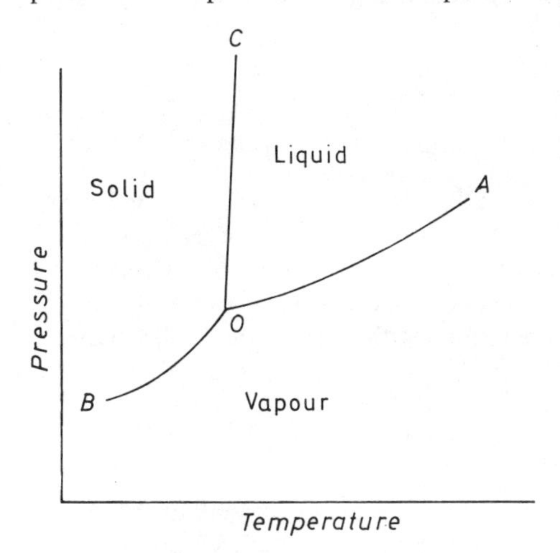

Figure 6.1 Phase diagram

Figure 6.1 (called a phase diagram) shows a plot of vapour pressure against temperature for a single substance. The curves *OA*, *OB* and *OC* represent simultaneous values of pressure and temperature at which two phases may be present at equilibrium. Hence, the curves *OA* and *OB* represent liquid and solid phases, respectively, in equilibrium with vapour, whereas curve *OC* represents solid in equilibrium with liquid, i.e., *OC* shows the effect of pressure on melting point. Point *O* (known as the TRIPLE POINT) represents the only set of conditions under which all three phases can co-exist (all three phases are in equilibrium).

6.2 Binary Solutions

6.2.1 SOLUTIONS OF GASES IN LIQUIDS

The solubility of a gas in a liquid is usually expressed as an ABSORPTION COEFFICIENT defined as the volume of gas, reduced to s.t.p., dissolved by unit volume of solvent at the temperature of the experiment under a partial pressure of the gas of 1 atmosphere.

When gases dissolve in a solvent, generally there is a liberation of heat, and so from Le Chatelier's principle (*see* Chapter 7), an increase of temperature results in a decrease in solubility of the gas. The most important factor affecting the solubility of a gas is pressure, and this is summarized by HENRY'S LAW which states that the mass of gas dissolved by a given volume of solvent, at constant temperature, is proportional to the pressure of the gas with which it is in equilibrium. Deviations from Henry's law occur when the gas departs from ideal behaviour (deviations would be expected at low temperatures and high pressures, particularly with easily liquefiable gases) or when there is a gas–solvent reaction, e.g., ammonia–water system shows considerable deviations from the law because of the reaction:

$$NH_3 + H_2O \rightleftharpoons NH_4^+ + OH^-$$

For the same reasons, gases such as hydrogen chloride and sulphur dioxide, etc., also show such deviations. If allowance is made for these reactions, Henry's law can be observed.

For a mixture of gases in equilibrium with a solvent, each gas dissolves according to its own partial pressure.

6.2.2 SOLUTIONS OF SOLIDS IN LIQUIDS

Certain physical properties of solutions are due to the number of particles of solute present in a given volume of solvent, and not on their nature. These physical properties are called COLLIGATIVE properties (some of which are described below); they offer ways of determining the relative molecular masses of dissolved solutes.

6.2.2.1 Lowering of Vapour Pressure

If a non-volatile solute is dissolved in a solvent, the vapour pressure of the solvent is lowered. The relationship between this lowering and concentration is given by RAOULT'S LAW, which says that the relative lowering of the vapour pressure is equal to the mole fraction of solute, or

$$\frac{P^0 - P}{P^0} = \frac{n}{n+N} \tag{6.1}$$

where P^0 and P = vapour pressure of pure solvent and solution, respectively, n and N = number of moles of solute and solvent respectively. If very dilute solutions are used (n is small), then:

$$\frac{P^0 - P}{P^0} \approx \frac{n}{N} \tag{6.2}$$

A solution which obeys Raoult's law is said to be an IDEAL SOLUTION (and ideality is approached as the solution is made more dilute).

Example: When 10 g of a non-volatile solute, A, were dissolved in 80 g of solvent X (of relative molecular mass 75 and vapour pressure, at a fixed temperature, of 500 mmHg), the vapour pressure of X was lowered to 490 mmHg. Calculate the relative molecular mass, m, of solute A.

From equation (6.1),

$$\frac{500 - 490}{500} = \frac{10/m}{(10/m) + (80/75)} = \frac{1}{50}$$

Hence $m = 459.4$

6.2.2.2 *Elevation of Boiling Point*

A liquid boils when its vapour pressure is equal to the atmospheric pressure. Since a dissolved, non-volatile solute lowers the vapour pressure of the solvent, the solution therefore needs to be heated to a higher temperature before its vapour pressure equals atmospheric pressure. Hence the boiling point is elevated

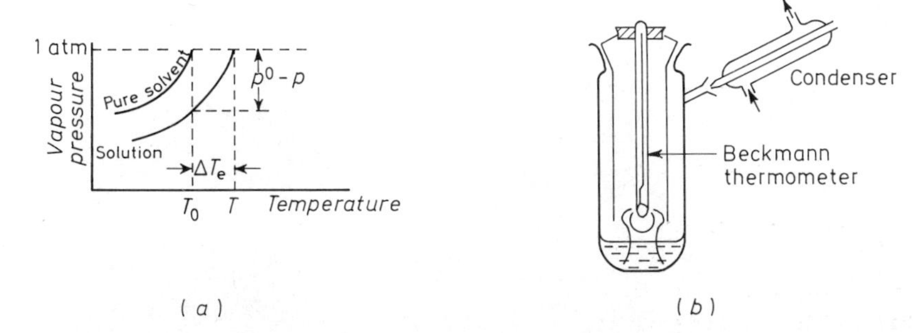

Figure 6.2 (a) Relationship of vapour pressure depression and boiling point elevation for a solution, and (b) apparatus for determination of elevation of boiling point

[from T_0 to T, *Figure 6.2(a)*]. The elevation of the boiling point, ΔT_e, is proportional to the MOLALITY of the solution (this can be shown from thermodynamic arguments), where the molality of a solution is defined as the number of moles of solute dissolved in 1000 g of solvent. Hence,

$$\Delta T_e = K_e M \tag{6.3}$$

where K_e is the proportionality constant called the MOLECULAR ELEVATION CONSTANT or EBULLIOSCOPIC CONSTANT, and M is the molality of the solution. If W g solvent contain n moles solute (or w/m moles solute, where w = mass of solute and m is its relative molecular mass) then 1000 g solvent would contain $[(w/m) \times (1000/W)]$ moles solute, and this is equal to the molality of the solution, M. On substitution into equation (6.3),

$$\Delta T_e = \frac{K_e \times 1000w}{mW} \tag{6.4}$$

or

$$m = \frac{K_e \times 1000w}{W\Delta T_e} \tag{6.5}$$

If one mole of solute is present ($w = m$) in 1000 g solvent ($W = 1000$) then $\Delta T_e = K_e$, or K_e is the elevation of the boiling point when one mole of any solute is dissolved in 1000 g of solvent. Since the theory only holds for dilute solutions, ΔT_e is found for a dilute solution and K_e is then found by simple proportion.

Boiling point elevations can be experimentally determined using Cottrell's boiling point apparatus [*Figure 6.2(b)*]. Here, bubbles of vapour carry liquid up through the jets, spraying it onto the bulb of the Beckmann thermometer (a very accurate thermometer which will read to 0.001 of a degree). The experiment is carried out using pure solvent and is then repeated when a weighed amount of solute has been added to the solvent; hence the elevation of boiling point is obtained by difference.

Example: The addition of 1 g of solute X to 50 g ethoxyethane (diethyl ether) caused its boiling point to be raised by 0.60 °C. Calculate the relative molecular mass of X (the molecular elevation constant of ethoxyethane is 2.02 °C kg^{-1} and its relative molecular mass is 74).

From equation (6.5),

$$m = \frac{2.02 \times 1000 \times 1}{50 \times 0.6} = \frac{2020}{30} = 67.3$$

6.2.2.3 Depression of Freezing Point

Figure 6.3(a) shows the vapour pressure curves for the solid and liquid states of the same substance. The point of intersection, T_0, is the melting or freezing point of the substance, i.e., the temperature at which solid and liquid are in equilibrium. If a solute is then dissolved in the liquid, the vapour pressure is lowered and hence the point of intersection, T, of the curves occurs at a lower temperature, giving a depression of freezing or melting point, ΔT_f. It can be shown, from thermodynamics, that ΔT_f is proportional to the molal concentration of the solution (provided dilute solutions are considered). Consequently, the same equation applies as for the elevation of boiling point, except in this case the constant, K_f [*see* equation (6.6)], is known as the CRYOSCOPIC CONSTANT or MOLECULAR DEPRESSION CONSTANT:

$$m = \frac{K_f \times 1000w}{W\Delta T_f} \tag{6.6}$$

Hence, K_f is the depression of the freezing (or melting) point when 1 mole of any solute is dissolved in 1000 g of the solvent (again, calculated by simple proportion from results for a dilute solution).

Freezing point depression (for relative molecular mass determination) can be determined by the Beckmann method [*Figure 6.3(b)*]. A known mass of solvent is placed in the inner tube and its freezing point is measured (supercooling

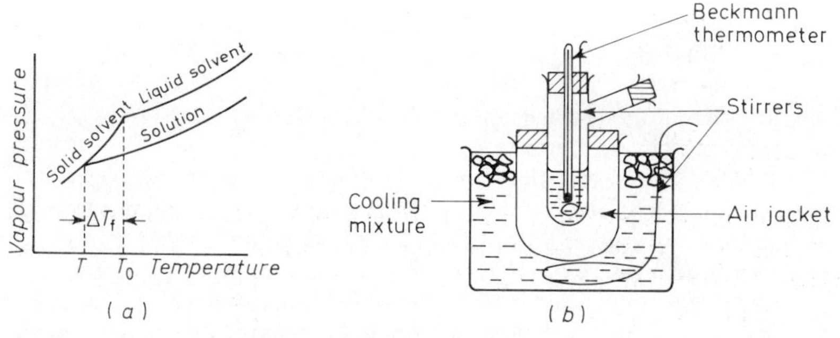

Figure 6.3 (a) Relationship of vapour pressure depression to freezing point depression for a solution, and (b) apparatus for determination of depression of freezing point

is kept to a minimum by cooling *via* an air jacket). The inner tube is removed until the solvent melts, when a weighed amount of solute is introduced. When dissolved, the freezing point of the solution is then measured – hence the depression of freezing point is known.

For measuring relative molecular masses of organic compounds, Rast's method is particularly useful. This involves fusing known masses of camphor and the unknown solute, resolidifying and then powdering; the melting point depression of the camphor is then determined. Camphor is favoured as the solvent since its cryoscopic constant is large and consequently a thermometer graduated in 0.1 °C intervals can be used.

6.2.2.4 Osmosis

Osmosis may be defined as the passage of solvent but not solute through a SEMIPERMEABLE MEMBRANE from solvent to solution or from a solution of low to one of higher concentration. A semipermeable membrane is one which allows the passage of solvent but not solute – examples include parchment, cellophane or other cellulosic materials, and copper(II) hexacyanoferrate(II) ($Cu_2[Fe(CN)_6]$) precipitated in a porous pot. Many plant and animal cells are surrounded by semipermeable membranes.

Osmosis can be demonstrated by the apparatus shown in *Figure 6.4(a)*. Initially, the levels of the solution in the column and the water in the beaker are the same. If left to stand, however, the level in the column rises until eventually

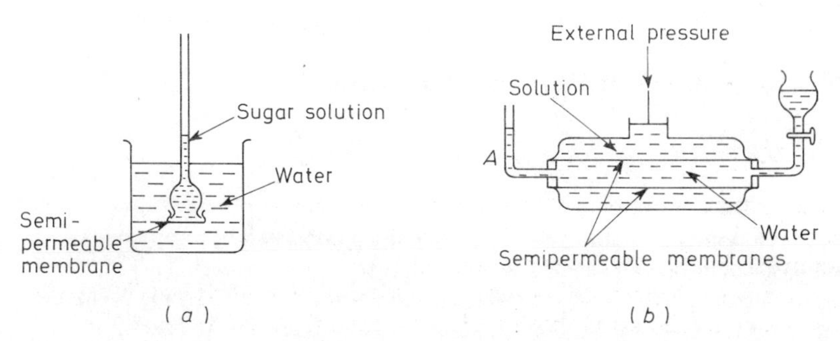

Figure 6.4 Osmosis

the hydrostatic pressure of the column balances the tendency for any excess water to pass through the membrane. A position of dynamic equilibrium has been reached since the rate of inflow of water (because of the tendency of the water to dilute the solution) equals the rate of outflow of water (because of the effect of the hydrostatic pressure).

The OSMOTIC PRESSURE, Π, of a solution is the external pressure which must be applied to it in order to prevent passage into it of excess solvent through a semipermeable membrane. In *Figure 6.4(a)*, the hydrostatic pressure is not equal to the osmotic pressure of the original sugar solution, since osmosis and hence dilution has already occurred. A more accurate measurement of Π is shown in *Figure 6.4(b)*, where the pressure needed to maintain the water level in side tube, A, is measured – hence osmosis is prevented and gives a value of Π for the solution used. Note that solutions which have equal osmotic pressures are called ISOTONIC solutions.

Two laws of osmosis have been observed:

(1) At a given temperature, the osmotic pressure, Π, of a dilute solution is proportional to its concentration, c,
(2) The osmotic pressure of a given solution is directly proportional to the absolute temperature, T.

Combining these laws gives

$$\Pi \propto cT$$

or

$$\Pi = \text{constant} \times cT$$

Since

$$c = \frac{\text{number of moles solute } (n)}{\text{Volume of solvent } (V)}$$

then, on substitution

$$\Pi V = \text{constant} \times n \times T$$

The constant is found to have the value of R, the gas constant. Hence:

$$\Pi V = nRT$$ (compare with ideal gas equation, $PV = nRT$ – Chapter 4, p. 29)

Since $n = w/m$, where w = mass of solute and m is its relative molecular mass, then:

$$\Pi V = \frac{wRT}{m} \quad \text{or} \quad m = \frac{wRT}{\Pi V}$$

Hence, measurements of Π for solutions of known concentration can be used for relative molecular mass determinations, particularly for polymers. This is because polymer solutions only give very small depressions of freezing point and elevations of boiling point, so that measurements by these methods are not sufficiently sensitive.

6.2.2.5 Effects of Association and Dissociation of Solute on Colligative Properties

When ethanoic acid is dissolved in a non-polar solvent such as benzene, a relative molecular mass of 120 instead of 60 is measured. This implies that only half the expected number of molecules are present [since, from equation (6.5) for example, this would halve ΔT_e and hence double m], and is therefore experimental evidence for dimerization (*see* p. 26). For acids where this association is not complete, relative molecular masses measured will be somewhere between the expected and twice the expected values.

In the case of electrolytes, where partial or complete dissociation occurs, the observed relative molecular mass will be lower than the true value because of the extra particles present. Van't Hoff introduced the factor, i, given by

$$i = \frac{\text{observed colligative property}}{\text{calculated colligative property (assuming no dissociation)}}$$

For a solution of, say, sodium chloride in water, dissociation can be complete and i takes the value of 2 – hence the calculated relative molecular mass will be half the true value. For sodium sulphate (Na_2SO_4), i will approach 3 giving a relative molecular mass of approximately one third of the true value.

6.2.2.6 Distribution of a Solute Between Two Immiscible Solvents

If a solute is added to a pair of immiscible solvents, e.g., ethoxyethane and water, then the ratio of the concentrations of the solute in the two layers when equilibrium has been attained is a constant at a fixed temperature. This is known as the DISTRIBUTION or PARTITION LAW and is written:

$$\frac{\text{Concentration of solute in solvent A}}{\text{Concentration of solute in solvent B}} = K$$

where K is known as the PARTITION or DISTRIBUTION COEFFICIENT. This law holds provided the solute is in the same state in both solvents. If association or dissociation of the solute occurs in one of the solvents, the law needs modification.

An important application of this distribution law is the purification of many compounds by solvent extraction, and is particularly useful in organic preparative chemistry (*see* Chapter 14, p. 142).

6.2.3 SOLUTIONS OF LIQUIDS IN LIQUIDS

On mixing two completely miscible ideal liquids, A and B (i.e., two liquids which show no heat change when mixed and obey Raoult's law over the whole composition range), the vapour pressure of the solution is simply the sum of the partial pressures of each component. *Figure 6.5(a)* shows a plot of the partial pressure of each liquid against its mole fraction, x, in the mixture, together with the total pressure of the solution. Mixtures are known which behave in a nearly ideal manner and give straight-line, or nearly straight-line, plots of vapour pressure against composition; an example is hexane and heptane at 30 °C. Most mixtures, however, show deviations from ideal behaviour (*see Figures 6.6* and *6.7*).

Figure 6.5(b) shows a boiling point or temperature–composition diagram for a near-ideal liquid mixture. The vapour curve shows the composition of the vapour in equilibrium with a liquid mixture at its boiling point, and shows how the vapour always contains more of the volatile component (in this case A). For example, if a liquid mixture of composition X is boiled, then the vapour has composition Y. If this is condensed and reboiled, vapour of composition Z is obtained, and repetition of this process will eventually give pure liquid A.

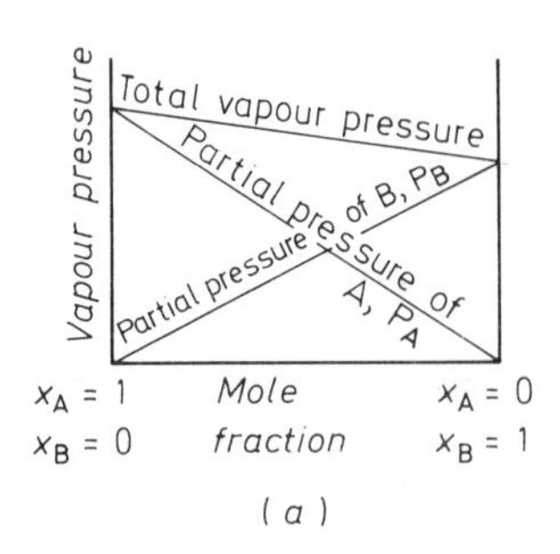

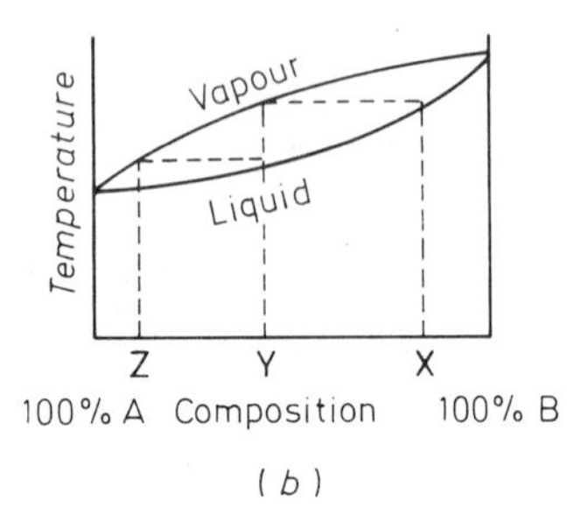

Figure 6.5 Vapour pressure and composition for mixtures of two volatile substances: (a) ideal, (b) near-ideal

Instead of separating the two liquids by several distillation processes, they are combined in one operation by using a FRACTIONATING COLUMN, which usually consists of a column packed with, for example, glass beads. This allows successive condensations and re-evaporations on the beads to occur as the vapour ascends the column, and separation is achieved in a single FRACTIONAL DISTILLATION.

6.2.3.1 *Deviations from Raoult's Law*

A non-ideal solution is formed when the process of mixing its components is accompanied by the evolution or absorption of heat. For example, if trichloromethane (chloroform) and propanone (acetone) are mixed, heat is evolved which implies that in solution the components are in a lower energy state than when pure. This is because hydrogen bonding occurs between unlike molecules, resulting in strong intermolecular attractions, as shown in structure (1). Hence, the

$$Cl_3C{-}H\cdots\cdots O{=}C(CH_3)_2$$

(1)

vapour pressure of each component is lower than would be predicted by Raoult's law, and NEGATIVE DEVIATION from the law is said to occur [*see Figure 6.6(a)*]. In such cases the vapour pressure curve may have a minimum value for a particular composition, and consequently the boiling point curve has a maximum [point X, *Figure 6.6(b)*]. At such a point, X, the liquid and vapour have

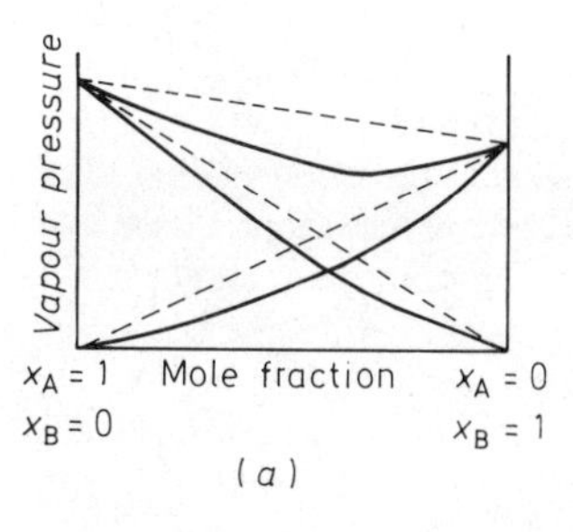

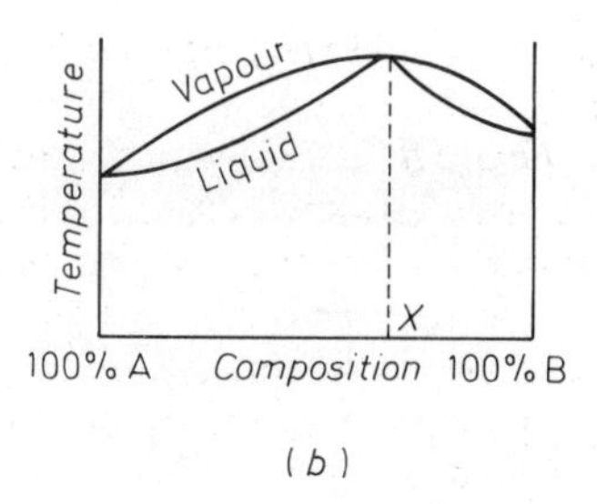

Figure 6.6 Negative deviations from Raoult's law, leading to a mixture of maximum boiling point

the same composition, i.e., the liquid boils at a constant temperature and will distil over without change of composition – such a mixture is called an AZEOTROPIC or CONSTANT BOILING POINT MIXTURE. An aqueous solution of hydrochloric acid behaves like this.

If on mixing the two liquids, absorption of heat occurs then the component molecules in solution are in a higher energy state than when pure, i.e., attractive forces between unlike molecules are weaker than those between molecules of the same kind. Hence, the escaping tendency of solution molecules is increased and so the vapour pressure of each component is higher than would be predicted

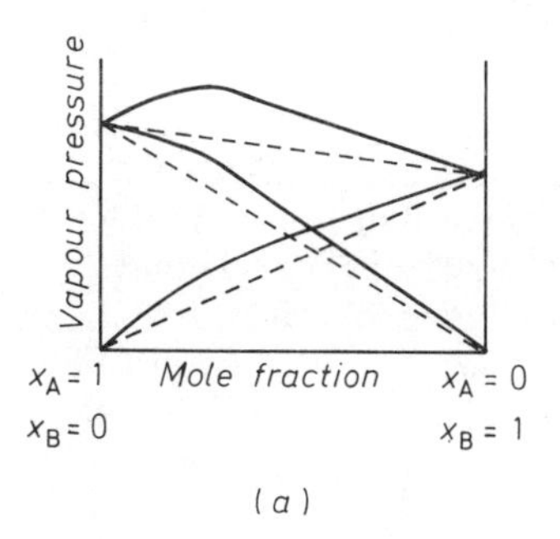

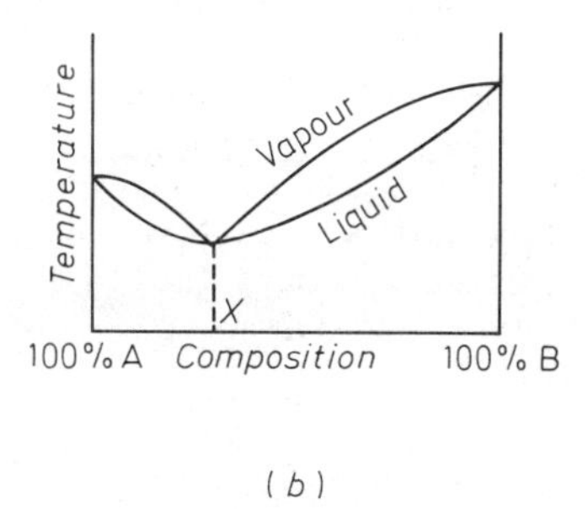

Figure 6.7 Positive deviations from Raoult's law, leading to a mixture of minimum boiling point

by Raoult's law. Solutions of this kind show POSITIVE DEVIATION from Raoult's law and this time the vapour pressure curve may show a maximum value with a consequent minimum value in the boiling point curve [*Figures 6.7(a)* and *6.7(b)*]. At the point *X* [*Figure 6.7(b)*] liquid and vapour again have the same composition and give a constant boiling mixture. Ethanol–water systems can behave like this.

6.2.4 STEAM DISTILLATION

Quite often in preparative chemistry, the desired product in the reaction flask is contaminated with other products, e.g., in the preparation of phenylamine (aniline) from nitrobenzene (p. 183), phenylamine is mixed with water, oxidation products of the reducing agent, and others. Although the boiling point of

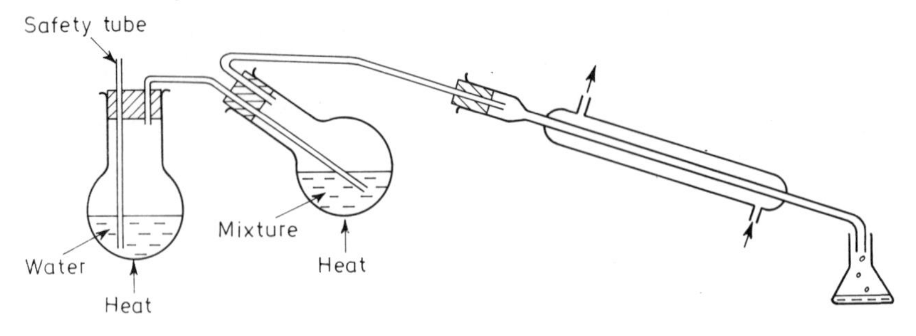

Figure 6.8 Apparatus for steam distillation

phenylamine is 184.6 °C, it can be recovered from the mixture, together with water, by passing steam through it; this is known as STEAM DISTILLATION (*see Figure 6.8*). In a mixture of two immiscible liquids, such as phenylamine and water, the vapour pressure of the system is equal to the sum of the two separate vapour pressures. Hence the mixture boils and distils over at a temperature below the boiling point of either constituent.

6.2.5 SIMPLE EUTECTIC MIXTURES

Figure 6.9 shows the effect of a substance B on the melting point of a substance A, and *vice versa*. Pure A has a melting point T_A °C and addition of B lowers its melting point, curve *AE* being followed as the amount of B increases. Addition of A to B gives the melting point curve *BE*. Point *E* shows a minimum point where both curves meet, and this 'easily melting' mixture is known as the EUTECTIC MIXTURE, the temperature at which it melts being the EUTECTIC TEMPERATURE (T_e °C). T_e °C is the lowest melting point of an A/B mixture.

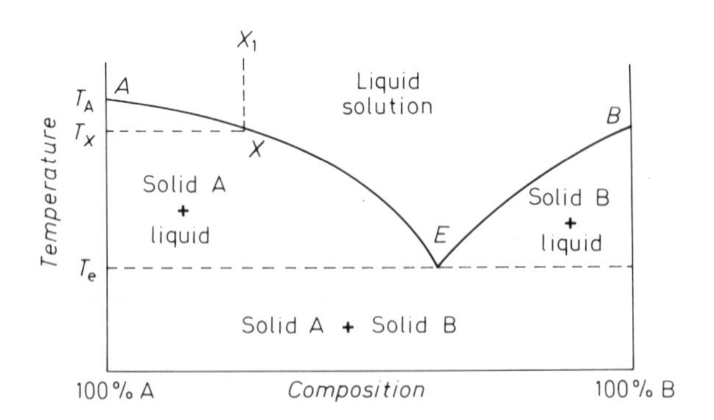

Figure 6.9 Eutectic curve

All mixtures above the curve exist in the molten state. If a molten mixture at point X_1 is cooled, pure A starts to separate out at point *X* (at temperature T_X °C), the liquid remaining then becoming richer in B and hence depressing the melting point further. The curve XE is now followed until point E is reached. At point E, A and B crystallize in the proportion corresponding to the eutectic mixture.

Examples of mixtures forming simple eutectics include lead–tin (common solder) and zinc–cadmium.

7

CHEMICAL EQUILIBRIA

7.1 Equilbrium

In Chapter 6 it was seen how a liquid, in a closed container, could reach a position of dynamic equilibrium as the rates of evaporation and condensation became equal. Chemical reactions, like phase changes, can be reversible. For example, ethanoic (acetic) acid and ethanol react to give ethyl ethanoate (acetate) and water and, similarly, the products can react to give the alcohol and acid:

$$CH_3COOH + C_2H_5OH \rightleftharpoons CH_3COOC_2H_5 + H_2O$$

Since, at any particular temperature, the rate of a reaction is proportional to the concentration of the reacting species (called the LAW OF MASS ACTION) the rate of forward reaction will, initially, be fast and will gradually slow down as the concentration of reactants decreases; conversely, the concentration of products will be increasing and so the rate of back reaction will gradually increase. At some stage, the rate of forward and back reactions will become equal and a dynamic equilibrium has been reached.

Consider the general reaction,

$$A + B \rightleftharpoons C + D$$

Rate of forward reaction,

$$r_1 \propto [A]\,[B] \quad \text{(where [ ] = molar concentration)}$$

or

$$r_1 = k_1 [A]\,[B]$$

Similarly, rate of back reaction,

$$r_2 \propto [C]\,[D]$$

or

$$r_2 = k_2 [C]\,[D]$$

(where k_1 and k_2 are proportionality constants for forward and back reactions respectively). Equilibrium is attained when

$$r_1 = r_2$$

or

$$k_1 [A]\,[B] = k_2 [C]\,[D]$$

or

$$\frac{[C]\,[D]}{[A]\,[B]} = \frac{k_1}{k_2} = K_c \qquad (7.1)$$

where K_c is the EQUILIBRIUM CONSTANT (in terms of molar concentrations at equilibrium) for the system and is constant at a particular temperature. For gaseous reactions, when the concentrations are in terms of partial pressures, the symbol K_p is used.

For the more general reaction

$$aA + bB \rightleftharpoons cC + dD$$

then equation (7.1) becomes:

$$K_c = \frac{[C]^c[D]^d}{[A]^a[B]^b} \qquad (7.2)$$

Equation (7.2) is known as the EQUILIBRIUM LAW*. The equilibrium constant can be calculated for a given reaction at a particular temperature once the equilibrium concentrations of the mixture are known.

Example: A mixture of 207 g of propanol and 207 g of ethanoic acid was kept at constant temperature until equilibrium was attained; 60 g of the alcohol remained. Calculate the equilibrium constant at this temperature (C = 12, H = 1, O = 16).

It is seen that both reactants have relative molecular masses of 60, so 207 g is 3.45 mol. Let the total volume of mixture be V litres†.

	CH_3COOH +	C_3H_7OH ⇌	$CH_3COOC_3H_7$ +	H_2O
Initial concn./mol l^{-1}	$3.45/V$	$3.45/V$	0	0
Equilibrium concn./mol l^{-1}	$1/V$	$1/V$	$2.45/V$	$2.45/V$

$$K_c = \frac{[CH_3COOC_3H_7][H_2O]}{[CH_3COOH][C_3H_7OH]} = \frac{(2.45/V) \times (2.45/V)}{(1/V) \times (1/V)} = 6.003$$

Example: Suppose that 3 moles of ethanoic acid and 1 mole of propanol are mixed at the same temperature as in above example. How many moles of ester will be formed at equilibrium? Let total volume be V litres, and let x moles of ester, and therefore x moles of water, be formed at equilibrium.

	CH_3COOH +	C_3H_7OH ⇌	$CH_3COOC_3H_7$ +	H_2O
Initial concn./mol l^{-1}	$3/V$	$1/V$	0	0
Equilibrium concn./mol l^{-1}	$(3-x)/V$	$(1-x)/V$	x/V	x/V

Now,

$$\frac{[CH_3COOC_3H_7][H_2O]}{[CH_3COOH][C_3H_7OH]} = \frac{(x/V) \times (x/V)}{[(3-x)/V] \times [(1-x)/V]} = 6 \text{ (from above)}$$

i.e.,

$$\frac{x^2}{3 - 4x + x^2} = 6 \quad \text{or} \quad 5x^2 - 24x + 18 = 0$$

*Note that the above derivation is not a general proof of the equilibrium law since it is based on a special type of rate equation which is not always obeyed. The equilibrium law can be derived from thermodynamic principles beyond the scope of this text.

†In this book we have continued to use as volume unit the litre (l). Elsewhere the cubic decimetre (dm^3) may increasingly be found. For practical purposes the two are the same, and molarity can be written as mol l^{-1} or mol dm^{-3}.

$$\text{Since } x = \frac{-b \pm \sqrt{b^2 - 4ac}}{2a}, \text{ then}$$

$$x = -\frac{(-24) \pm \sqrt{(-24)^2 - 360}}{10} = 3.87 \text{ or } 0.93$$

But $x = 3.87$ is impossible, so moles of ester formed = 0.93. (There will also be 0.93 mole water, 0.07 mole propanol and 2.07 moles ethanoic acid at equilibrium.)

7.2 Heterogeneous Equilibria

The systems considered above are homogeneous systems (i.e., reactants and products are in the same phase). However, the equilibrium law can be applied to heterogeneous systems, as shown in the following example.

If calcium carbonate is heated in an open vessel and the carbon dioxide is swept away, complete conversion from carbonate to oxide can occur. If, however, decomposition occurs in a closed vessel, then the pressure of carbon dioxide increases and back reaction can occur:

$$CaCO_{3\,(s)} \rightleftharpoons CaO_{(s)} + CO_{2\,(g)}$$

At equilibrium, in terms of partial pressures,

$$\frac{(P_{CaO}) \times (P_{CO_2})}{(P_{CaCO_3})} = \text{constant}$$

and since at any given temperature P_{CaO} and P_{CaCO_3} are constant, then

$$\text{constant} = \frac{\text{constant} \times P_{CO_2}}{\text{constant}}$$

or

$$K_p = P_{CO_2}$$

At any given temperature, therefore, the partial pressure of carbon dioxide is constant and does not depend on the amounts of calcium carbonate or oxide present (provided some of each solid is present).

Similarly, for the dissociation of ammonium carbamate:

$$NH_4CO_2NH_{2\,(s)} \rightleftharpoons 2NH_{3\,(g)} + CO_{2\,(g)}$$

$P_{NH_4CO_2NH_2}$ is a constant at a given temperature, and so

$$K_p = (P_{NH_3})^2 \times (P_{CO_2})$$

7.3 Le Chatelier's Principle and Its Applications

Le Chatelier's principle states that if the conditions of a system at equilibrium are changed, the system moves in such a way as to oppose the effects of the change. Consequently, if the concentration of the reactants is increased, the position of equilibrium moves to the right (or forwards) since in this way some of the reactants will be used up and so their concentration falls. Similarly, increasing the concentration of products shifts the equilibrium position to the left (or backwards). However, the value of the equilibrium constant stays the same, provided the temperature is kept constant.

In the case of an exothermic reaction (where heat can be considered as a product), increasing the temperature results in the system trying to use up the extra heat and so the equilibrium moves backwards or to the left. If, however, the temperature is lowered the system responds by the equilibrium moving forwards or to the right, hence producing more heat. Using similar arguments, it can be seen that for an endothermic reaction, an increase of temperature moves the equilibrium to the right and a decrease of temperature moves the equilibrium to the left.

For gaseous reactions, if the number of moles of gaseous products exceeds the number of moles of gaseous reactants, then the left hand side of the equation is considered to be the low pressure side, and the right is the high pressure side. Hence, increasing the pressure results in the equilibrium moving to the left to reduce the pressure, and *vice versa*. However, if the left hand side of the equation is the high pressure side, increasing the pressure shifts the position of equilibrium forwards (to reduce the pressure), and *vice versa*. If the number of moles of gaseous reactants and products are the same, then pressure changes have no effect on the position of equilibrium.

The following examples illustrate the above principles.

7.3.1 HABER PROCESS (Industrial preparation of ammonia)

$$N_2 + 3H_2 \rightleftharpoons 2NH_3; \ \Delta H^{\ominus}_{298} = -46 \text{ kJ mol}^{-1}$$

Since the right hand side of the equation is the low pressure side, a high pressure should favour a greater yield of ammonia; hence the process is run at approximately 200 atmospheres.

As the reaction is exothermic, a low temperature should shift the equilibrium to the right resulting in more ammonia being produced. However, this results in a slow rate of attainment of equilibrium (since, in general, reaction rate increases with temperature), so a catalyst (finely divided iron) is used at a compromise temperature of about 550 °C.

7.3.2 CONTACT PROCESS (Manufacture of sulphuric acid)

$$2SO_2 + O_2 \rightleftharpoons 2SO_3; \ \Delta H^{\ominus}_{298} = -96 \text{ kJ mol}^{-1}$$

As the process is exothermic, temperature considerations are similar to those of the Haber process; hence a catalyst of vanadium(V) oxide (vanadium pentoxide) is used at compromise temperatures of up to 450 °C.

As in the Haber process, high pressure would be expected to increase the yield, but in practice, the process runs satisfactorily at near-atmospheric pressures.

8

IONIC EQUILIBRIA

8.1 Solubility Product

In a saturated solution of silver chloride, a dynamic equilibrium exists between solid silver chloride and its ions in solution, i.e., the rate at which ions leave the crystal lattice and enter solution equals the rate at which they leave solution and re-enter the crystal lattice. Therefore:

$$AgCl_{(s)} \rightleftharpoons Ag^{+}_{(aq)} + Cl^{-}_{(aq)}$$

From the equilibrium law:

$$\frac{[Ag^{+}][Cl^{-}]}{[AgCl]} = \text{constant (at constant temperature)}$$

Since [AgCl] is a constant, then

$$K_s = [Ag^{+}][Cl^{-}]$$

where K_s is a new constant, called the SOLUBILITY PRODUCT and has the value of 2×10^{-10} mol^2 l^{-2} at 25 °C for AgCl. Precipitation occurs when this value is exceeded. If the Cl^- concentration is increased by adding, for example, NaCl to the solution, the above equilibrium is shifted to the left, causing precipitation of silver chloride and therefore reducing the Ag^+ concentration – hence the K_s value is restored. Some examples of K_s for other systems are given in *Table 8.1*.

Table 8.1 Solubility products

System	K_s	*Value* (298 K)
$CaSO_{4(s)} \rightleftharpoons Ca^{2+}_{(aq)} + SO^{2-}_{4(aq)}$	$[Ca^{2+}][SO_4^{2-}]$	2.0×10^{-5} mol^2 l^{-2}
$Ag_2CrO_{4(s)} \rightleftharpoons 2Ag^{+}_{(aq)} + CrO^{2-}_{4(aq)}$	$[Ag^{+}]^2[CrO_4^{2-}]$	3.0×10^{-12} mol^3 l^{-3}
$Fe(OH)_{3(s)} \rightleftharpoons Fe^{3+}_{(aq)} + 3OH^{-}_{(aq)}$	$[Fe^{3+}][OH^{-}]^3$	8.0×10^{-40} mol^4 l^{-4}

Solubility product is only applicable to solutions of sparingly soluble salts, since for more soluble salts interionic attractions may occur (leading to non-ideality).

Example: The solubility of magnesium hydroxide is 9.86×10^{-3} g l^{-1} at 25 °C. Calculate its solubility product K_s (Mg = 24, O = 16, H = 1).

Since the relative molecular mass of $Mg(OH)_2$ = 58, its solubility is

$$9.86 \times 10^{-3}/58 = 1.7 \times 10^{-4} \text{ mol l}^{-1}.$$

	$Mg(OH)_{2\,(aq)}$	$\rightarrow$	$Mg^{2+}_{(aq)}$	$+\ 2OH^-_{(aq)}$
$10^4 \times$ Concn./mol l^{-1}	1.7		1.7	3.4

$$K_s = [Mg^{2+}][OH^-]^2 = 1.7 \times 10^{-4} \times (3.4 \times 10^{-4})^2$$
$$= 1.96 \times 10^{-11} \text{ mol}^3 \text{ l}^{-3}$$

8.1.1 COMMON ION EFFECT

If concentrated hydrochloric acid is added to a saturated solution of common salt, some of the salt precipitates. This is because the fully dissociated hydrochloric acid increases the Cl^- concentration and drives the equilibrium (below) to the left.

$$NaCl_{(s)} \rightleftharpoons Na^+_{(aq)} + Cl^-_{(aq)}$$

This is called the COMMON ION EFFECT. Similarly, addition of ammonium chloride to ammonium hydroxide suppresses the dissociation of the latter because of the increase in NH_4^+ concentration.

This effect can be industrially useful. For example, in the manufacture of soap (a sodium salt), addition of common salt causes the soap to precipitate because of the increase in Na^+ concentration (a process called 'salting out').

8.2 Acids and Bases

Brønsted and Lowry defined acids as proton donors, whilst bases were proton acceptors. The first ionization of sulphuric acid can be written:

$$\underset{\text{(Acid)}}{H_2SO_4} + \underset{\text{(Base)}}{H_2O} \rightleftharpoons \underset{\text{(Conjugate acid)}}{H_3O^+} + \underset{\text{(Conjugate base)}}{HSO_4^-}$$

Here, water is acting as a base by accepting a proton and is being converted into its CONJUGATE ACID, H_3O^+; the acid, H_2SO_4, donates the proton and is converted into its CONJUGATE BASE, HSO_4^-.

Lewis defined acids as electron pair acceptors, and bases as electron pair donors. For example, trimethylamine and boron trifluoride react to form a solid salt since the former has a lone pair of electrons whilst BF_3 is electron deficient because the boron (in the compound) only has six electrons in its outer shell.

$$\underset{\text{(Lewis base)}}{(CH_3)_3N\!:} + \underset{\text{(Lewis acid)}}{BF_3} \rightleftharpoons \underset{\text{(The salt)}}{(CH_3)_3\overset{+}{N}-\overset{-}{B}F_3}$$

When an acid, HA, is added to water, the following equilibrium is set up:

$$HA + H_2O \rightleftharpoons H_3O^+ + A^-$$

The strength of the acid is determined by the position of this equilibrium, which is well to the right for strong acids such as HCl, HNO_3, etc. (hence the conjugate base A^- must be a weak base), and is well to the left for weak acids such as ethanoic acid, CH_3COOH (hence A^- must be a strong base). It should be noted, however, that acids only behave as acids when there is a base present to accept the proton, i.e., HCl is a strong acid in water but not in benzene.

Consider one mole of a weak acid, HA, in V litres of solution. Let x be its degree of dissociation.

	HA	⇌	H^+	+	A^-
Initial concn./mol l^{-1}	$1/V$		0		0
Equilibrium concn./mol l^{-1}	$(1-x)/V$		x/V		x/V

From the equilibrium law:

$$\frac{[H^+][A^-]}{[HA]} = \text{constant (at constant temperature)}$$

Hence:

$$\frac{\frac{x}{V} \times \frac{x}{V}}{\frac{1-x}{V}} = K = \frac{x^2}{V^2} \times \frac{V}{1-x} = \frac{x^2}{V(1-x)} \qquad (8.1)$$

This is OSTWALD'S DILUTION LAW. K is called the DISSOCIATION or IONIZATION CONSTANT (often given the symbol K_a). If the acid is weak and hence x is small, then $(1-x) \approx 1$, and so

$$x^2/V = K \quad \text{or} \quad x = \sqrt{KV} \qquad (8.2)$$

Therefore, the degree of dissociation is directly proportional to the square root of the volume containing one mole of HA. Hence x varies with concentration (but K remains constant at constant temperature).

8.2.1 THE CONCEPT OF pH

Water, although mainly covalent and therefore a non-electrolyte, does dissociate slightly:

$$H_2O \rightleftharpoons H^+ + OH^-$$

or, more accurately,

$$2H_2O \rightleftharpoons H_3O^+ + OH^-$$

From the equilibrium law:

$$K_c = \frac{[H^+][OH^-]}{[H_2O]}$$

Since very little dissociation occurs, the water concentration may be regarded as constant, and so:

$$[H^+][OH^-] = K_c \times \text{constant} = K_w \qquad (8.3)$$

K_w is called the IONIC PRODUCT OF WATER and, as determined from conductivity measurements, has the value 10^{-14} mol^2 l^{-2} at 25 °C. Since one molecule of water gives one H^+ ion and one OH^- ion, we can write:

$$[H^+] = [OH^-] = 10^{-7} \text{ mol l}^{-1} \text{ at } 25\,°C$$

Hence, a neutral aqueous solution is one which contains 10^{-7} mol l^{-1} of H^+ ions (or OH^- ions). If acid is added to the water, the H^+ concentration is increased and since K_w is a constant, the OH^- concentration must decrease so that the K_w value is restored.

The acidity of a solution is expressed in terms of the H^+ concentration, but since this can involve negative powers of ten it is more convenient to use the term 'pH', defined as

$$pH = -\log_{10} [H^+] \tag{8.4}$$

For a neutral aqueous solution, therefore, where $[H^+] = 10^{-7}$ mol l^{-1}, pH = 7. Alkalinity can also be described by pH, since from equation (8.3):

$$[H^+] \times [OH^-] = 10^{-14}$$

Taking negative logarithms,

$$-\log_{10} [H^+] - \log_{10} [OH^-] = 14$$

or

$$pH + pOH = 14 \quad \text{or} \quad pH = 14 - pOH$$

Example (i): Calculate the pH of 0.005M-sulphuric acid (assuming complete dissociation).

	$H_2SO_{4\,(aq)}$	→ $2H^+_{(aq)}$	+ $SO^{2-}_{4\,(aq)}$
Concn/mol l^{-1}	0.005	0.01	0.005

$[H^+] = 0.01$ or 10^{-2} mol l^{-1}

hence

$$pH = -\log_{10}(10^{-2}) = 2$$

Example (ii): Calculate the pH of 0.01M-potassium hydroxide solution (assuming complete dissociation).

	$KOH_{(aq)}$	→ $K^+_{(aq)}$	+ $OH^-_{(aq)}$
Concn./mol l^{-1}	0.01	0.01	0.01

$[OH^-] = 10^{-2}$ mol l^{-1}

But

$$[H^+][OH^-] = 10^{-14} \text{ mol}^2 \text{ l}^{-2}$$

Hence

$$[H^+] = 10^{-12} \text{ mol l}^{-1}$$

Therefore

$$pH = -\log_{10}(10^{-12}) = 12$$

Alternative method:

$$[OH^-] = 10^{-2} \text{ mol l}^{-1}$$

Hence

$$pOH = -\log_{10}(10^{-2}) = 2$$

But

$$pH = 14 - pOH$$

Hence

$$pH = 14 - 2 = 12$$

Example (iii): Calculate the pH of 0.01 M-methanoic (formic) acid. ($K_a = 1.6 \times 10^{-4}$ mol l^{-1} at 25 °C)

Since this acid is not fully dissociated, let its degree of dissociation be x:

	$HCOOH$	$\rightleftharpoons$	$HCOO^-$	+	H^+
Equilibrium concn./mol l^{-1}	$(1-x)/V$		x/V		x/V

From equation (8.2), $x = \sqrt{1.6 \times 10^{-4} \times 10^2}$ (since the volume containing one mole of acid is 100 or 10^2).

Hence

$$x = \sqrt{1.6 \times 10^{-2}} = 1.26 \times 10^{-1} \text{ (or 0.126)}$$

But

$$[H^+] = x/V \text{ mol l}^{-1} \text{ (from the above equilibrium)}$$

Therefore

$$[H^+] = 0.126/10^2 = 1.26 \times 10^{-3} \text{ mol l}^{-1}$$

Hence

$$pH = -\log_{10}(1.26 \times 10^{-3}) = -(\bar{3}.1004)$$

$$\approx -(-2.9) \approx 2.9$$

8.2.2 STRENGTHS OF WEAK ACIDS AND BASES

For a weak acid, HA, in water, the following equilibrium is established:

$$HA + H_2O \rightleftharpoons H_3O^+ + A^-$$

The equilibrium constant is given by

$$\frac{[H_3O^+][A^-]}{[HA][H_2O]} = K_c$$

However, in dilute solution, $[H_2O]$ is large and effectively constant, and so:

$$\frac{[H_3O^+][A^-]}{[HA]} = K_c \times \text{constant} = K_a \qquad (8.5)$$

K_a is the acid dissociation constant and is a measure of the strength of the acid. Since the above equilibrium is to the left for a weak acid (*see* before), K_a can be very small and so it is more convenient to use pK_a, where p$K_a = -\log_{10}K_a$. The lower the pK_a, the stronger is the acid, as shown by the following set of values: CH_3COOH, pK_a 4.76; $CH_2ClCOOH$, pK_a 2.86; $CHCl_2COOH$, pK_a 1.29; CCl_3COOH, pK_a 0.65. The reasons for the increasing strength of these acids are outlined in Chapter 15, p. 176.

Example (iv) shows an alternative method for calculating the problem given in Example (iii).

Example (iv): Calculate the pH of 0.01M-methanoic (formic) acid ($K_a = 1.6 \times 10^{-4}$ mol l^{-1} at 25 °C).

Since from one molecule of HCOOH, one H_3O^+ ion and one $HCOO^-$ ion are generated, then $[H_3O^+] = [HCOO^-]$. Also, since the acid is weak, the initial and equilibrium concentrations of HCOOH will be approximately the same, since little dissociation occurs.

Therefore, from equation (8.5),

$$\frac{[H_3O^+]^2}{10^{-2}} = 1.6 \times 10^{-4}$$

Hence

$$[H_3O^+]^2 = 1.6 \times 10^{-6}$$

or

$$[H_3O^+] = 1.26 \times 10^{-3} \text{ mol l}^{-1}$$

Therefore

$$\text{pH} = -\log_{10}(1.26 \times 10^{-3}) = -(\bar{3}.1004) \approx 2.9$$

The strength of a base, B, in water may be estimated from the equilibrium:

$$B + H_2O \rightleftharpoons BH^+ + OH^-$$

Assuming that $[H_2O]$ is large and effectively constant, then

$$K_b = \frac{[BH^+][OH^-]}{[B]} \qquad (8.6)$$

K_b is the BASE DISSOCIATION CONSTANT and is a measure of the strength of the base. Since this equilibrium is to the left for a weak base, values of K_b can be small and so pK_b is used, where p$K_b = -\log_{10}K_b$. The smaller the value of pK_b, the stronger is the base. Some values are: NH_3, pK_b 4.75; CH_3NH_2, pK_b 3.36; $(CH_3)_2NH$, pK_b 3.23. The reasons for this order of increasing base strength are discussed in Chapter 15, p. 184.

The strength of a base may also be expressed in terms of pK_a (e.g., a weak base in its protonated form, BH^+, will tend to release the proton relatively easily). Hence,

$$BH^+ + H_2O \rightleftharpoons B + H_3O^+$$

and so

$$K_a = \frac{[B][H_3O^+]}{[BH^+]}$$

pK_b values may be converted into pK_a values by using the equation:

$$pK_a + pK_b = 14 \text{ (at 25 °C)}$$

It should be noted that in the action of water with weak acid, HA, the water accepted a proton, but with the weak base, B, water donated a proton. A material which can be both basic and acidic is termed AMPHIPROTIC.

Example (v): In an experiment to determine K_a for a weak acid HA, 25.00 cm^3 of the acid solution were neutralized by 24.52 cm^3 of 0.1M-sodium hydroxide solution. By separate experiment, the acid solution was found to have a pH of 2.80. Calculate K_a for the acid.

The neutralization can be represented by:

$$HA + NaOH \rightarrow NaA + H_2O \text{ (since acid is monobasic)}$$

Moles of alkali used for neutralization = (24.52/1000) × 0.1 = 0.002452 mol. Therefore, moles of HA used in neutralization also equals 0.002452 mol, and since this is contained in 25.00 cm^3 of solution the molarity of the acid solution is given by:

$$0.002452 \times 40 = 0.0981\text{M}$$

Hence the acid solution contains 0.0981 mol l^{-1}
Since pH of acid solution = 2.80, then

$$2.80 = -\log_{10}[H^+]$$

or

$$\log_{10}[H^+] = -2.80 = (\bar{3}.20)$$

Hence

$$[H^+] = 1.58 \times 10^{-3} \text{ mol l}^{-1}$$

From equation (8.5),

$$K_a = \frac{1.58 \times 10^{-3} \times 1.58 \times 10^{-3}}{0.0981} \quad \text{(since } [H^+] = [A^-])$$

$$= 2.55 \times 10^{-5} \text{ mol l}^{-1}$$

8.2.3 BUFFER SOLUTIONS

A buffer solution is a solution of a weak acid and one of its salts (or a weak base and one of its salts) whose pH does not alter appreciably when small or moderate quantities of strong acids or alkalis are added. A typical example would be a solution containing ethanoic acid and sodium ethanoate (acetate). Two situations can now be considered:

(1) Suppose some strong acid is added to this mixture. The strong acid will readily react with ethanoate ions (mainly from the dissociated salt) to form ethanoic acid. Since the latter is only slightly dissociated, the pH will be almost unchanged.
(2) Suppose some strong alkali is added. In this case, the ethanoic acid reacts with the alkali (to form salt and water), neutralizing its effect, and again the pH remains almost unchanged.

Example: Consider one litre of buffer solution containing one mole of ethanoic acid and one mole of sodium ethanoate. Calculate the pH of the buffer (K_a for ethanoic acid is 1.7×10^{-5} mol l^{-1} at 25 °C).
From equation (8.5),

$$K_a = \frac{[CH_3COO^-][H_3O^+]}{[CH_3COOH]}$$

In the mixture, the $[CH_3COO^-]$ will be high because of the dissociated salt. Hence, $[CH_3COO^-] \approx$ salt concentration (neglecting the small amount of CH_3COO^- ions from the acid dissociation) = 1 mol l^{-1}. Also, because the acid is weak, the initial and equilibrium concentrations of CH_3COOH will be approximately the same.
Hence

$$K_a \approx \frac{[\text{Salt}][H_3O^+]}{[\text{Acid}]} \quad \text{or} \quad 1.7 \times 10^{-5} = \frac{1 \times [H_3O^+]}{1}$$

Therefore

$$[H_3O^+] = 1.7 \times 10^{-5} \text{ mol l}^{-1}$$

so

$$\text{pH} = -(\bar{5}.2304) = 4.77$$

Example: 10 cm^3 of 1M-hydrochloric acid are added to one litre of the above buffer solution. Calculate the resulting pH.

$$CH_3COONa + HCl \rightarrow CH_3COOH + NaCl$$

10 cm^3 of 1M-HCl represents $(10/1000) \times 1 = 0.01$ mole HCl. This will react with 0.01 mole CH_3COONa (leaving 0.99 mol l^{-1} of salt) to form 0.01 mole of CH_3COOH (giving a total $[CH_3COOH]$ of 1.01 mol l^{-1}).
Since

$$K_a \approx \frac{[\text{Salt}][H_3O^+]}{[\text{Acid}]}$$

then

$$[H_3O^+] = \frac{1.7 \times 10^{-5} \times 1.01}{0.99} = 1.7343 \times 10^{-5} \text{ mol l}^{-1}$$

Hence

$$\text{pH} = -(\bar{5}.2391) = -(-5 + 0.24) = 4.76$$

Hence, the pH has hardly changed from its initial value (of 4.77).

8.3 Salt Hydrolysis

When some salts are added to water, the resulting solution is not always neutral. This is because the salt undergoes HYDROLYSIS with the water. Many salts are hydrolysed; the hydrolysis which occurs from hydrated cations, e.g., $[Al(H_2O)_6]^{3+}$ or $[Fe(H_2O)_6]^{3+}$, is discussed in Chapters 12 and 13. Other systems are now outlined.

8.3.1 SALT OF WEAK ACID AND STRONG BASE, e.g., sodium ethanoate

$$H_2O \rightleftharpoons H^+ + OH^-$$

$$CH_3COONa \rightarrow CH_3COO^- + Na^+$$

$$\downarrow\uparrow$$

$$CH_3COOH$$

Hydrolysis results in the formation of ethanoic acid and sodium hydroxide. Since the former is only partially dissociated, whilst the latter is fully dissociated, the solution contains more OH^- ions than H^+ ions, and is therefore alkaline. Similarly, soluble carbonates and hydrogen carbonates, cyanides and sulphides also give alkaline solutions because the weak acids formed (H_2CO_3, HCN and H_2S) are only slightly dissociated.

8.3.2 SALT OF WEAK BASE AND STRONG ACID, e.g., ammonium chloride

$$H_2O \rightleftharpoons H^+ + OH^-$$

$$NH_4Cl \rightarrow Cl^- + NH_4^+$$

$$\downarrow\uparrow$$

$$NH_4OH$$

Since fully dissociated hydrochloric acid and partially dissociated base (NH_4OH) are produced, an acidic solution results.

8.3.3 SALT OF WEAK ACID AND WEAK BASE, e.g., ammonium ethanoate

$$H_2O \rightleftharpoons H^+ + OH^-$$

$$CH_3COONH_4 \rightarrow CH_3COO^- + NH_4^+$$

$$\downarrow\uparrow \qquad \downarrow\uparrow$$

$$CH_3COOH \qquad NH_4OH$$

In this situation, hydrolysis gives a weak acid and a weak base, and since both only partially dissociate to roughly the same extent, the resultant solution is approximately neutral.

8.3.4 SALT OF STRONG ACID AND STRONG BASE, e.g., sodium chloride

$$H_2O \rightleftharpoons H^+ + OH^-$$
$$NaCl \rightarrow Cl^- + Na^+$$

No weak acids or weak bases are formed with this system; hence the solution remains neutral.

8.4 Indicators and Acid–Base Titrations

An indicator is a weak acid or a weak base which changes colour when the pH of the solution is changed. To illustrate this action, consider methyl orange, which can be written as MeOH (weak base type indicator).

$$\underset{\text{(yellow)}}{MeOH} \rightleftharpoons \underset{\text{(red)}}{Me^+} + \underset{\text{(colourless)}}{OH^-}$$

If acid is added, the equilibrium moves to the right (as OH^- ions are removed by H^+ ions to produce water) – hence the red colour predominates. On adding alkali, the equilibrium moves to the left, and so the yellow colour is predominant.

Phenolphthalein, which is a weak acid type indicator, can be represented as HPh.

$$\underset{\text{(colourless)}}{HPh} \rightleftharpoons H^+ + \underset{\text{(pink)}}{Ph^-}$$

From the equilibrium, it can be seen that the indicator will be colourless under acid conditions and pink under alkaline conditions.

The pH range over which indicators change colour depends on their K_a or K_b values. *Table 8.2* gives the colour and pH ranges for some common indicators. In choosing an indicator for a particular titration, the type of titration has to be considered.

8.4.1 STRONG BASE–STRONG ACID

Consider a titration between 25 cm^3 0.1M-NaOH solution and 0.1M-HCl. When 24.95 cm^3 of HCl have been added, there are $(0.05/1000) \times 0.1$ mol $= 5 \times 10^{-6}$ mol of NaOH left in roughly 50 cm^3 solution; this is 10^{-4} mol l^{-1} of OH^-, i.e., solution has a pH of 10. When 25.05 cm^3 of HCl have been added there are $(0.05/1000) \times 0.1$ mol of HCl in roughly 50 cm^3 solution, which is 10^{-4} mol l^{-1} of H^+, i.e., pH is 4. So on adding only 0.1 cm^3 of acid, the pH has changed from 10 to 4 (*see* curve *AD, Figure 8.1*), and therefore any indicator in *Table 8.2* is suitable for this titration.

8.4.2 STRONG BASE–WEAK ACID

Curve *AC* (*Figure 8.1*) shows the type of curve obtained from, for example, a sodium hydroxide–ethanoic acid titration. An indicator which changes colour in the approximate range of pH 7 to 11 is therefore needed, i.e., phenolphthalein.

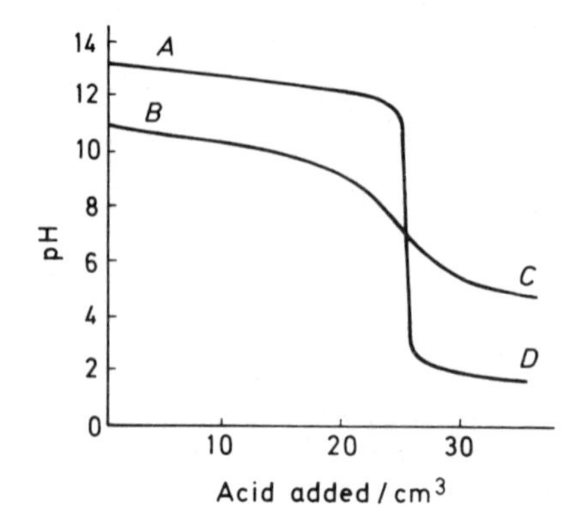

Figure 8.1 The pH *change during titrations of* 25 cm^3 0.1M-*base with* 0.1M-*acid*

8.4.3 WEAK BASE–STRONG ACID

In a titration of, say, ammonium hydroxide with hydrochloric acid, pH curve *BD* is followed (*Figure 8.1*). This time, an indicator which changes colour in the approximate range of pH 3 to 7 is needed, i.e., methyl orange or methyl red.

Table 8.2 Indicators

Indicator		*pH Range*		
Methyl orange	(orange)	3.1	4.5	(yellow)
Methyl red	(red)	4.7	6.4	(yellow)
Litmus	(red)	6.0	8.0	(blue)
Phenolphthalein	(colourless)	8.5	10.0	(red)

8.4.4 WEAK BASE–WEAK ACID

This would be the situation in a titration of ammonium hydroxide with ethanoic acid (curve *BC, Figure 8.1*). Since there is no sharp change in pH, there is no suitable indicator for this titration.

It should be noted that by careful choice of indicator, the various stages in a reaction can be identified. For example, consider the neutralization of sodium carbonate solution by dilute hydrochloric acid. The overall reaction is:

$$Na_2CO_3 + 2HCl \rightarrow 2NaCl + H_2O + CO_2$$

By using phenolphthalein indicator the reaction

$$Na_2CO_3 + HCl \rightarrow NaHCO_3 + NaCl$$

can be identified, since $NaHCO_3$ is a weak base (salt hydrolysis – *see* before). By then using methyl orange or methyl red, the titration can proceed further:

$$NaHCO_3 + HCl \rightarrow NaCl + H_2O + CO_2$$

8.5 Determination of pH

Two methods of pH determination will be described.

8.5.1 INDICATOR METHOD

Initially, the approximate pH of a given solution is found by means of universal indicator. Once known the actual pH can be found by comparing the colour of the solution in a specific indicator with the colour produced by this indicator in a buffer solution of known pH (the approximate determination tells us which range of buffers to use).

8.5.2 pH METER

This consists of a glass bulb (permeable to H^+ ions) containing 0.1M-HCl into which dips a platinum wire with its end plated with silver and then with silver chloride. When the electrode is placed in a solution of different pH, a potential difference is set up across the glass. By combining with a reference electrode (such as a calomel electrode – *see* Chapter 9) and measuring the e.m.f. produced using a valve voltmeter or other voltmeter of extremely high input impedance (calibrated in units of pH), the pH of the solution is determined.

9

ELECTROCHEMISTRY

9.1 Electrochemical Cells

When a metal is placed in a solution of its ions, a potential difference is set up, between the metal and solution, called the ELECTRODE POTENTIAL of the metal (and the metal/metal salt solution is known as a HALF CELL). This is because an equilibrium is established between the tendency of the metal to lose electrons and pass into solution as ions, and the opposing tendency for the ions in solution to gain electrons and be deposited on the metal. This results in the metal acquiring either a negative or positive charge depending on the position of the equilibrium:

$$M \rightleftharpoons M^{n+} + ne^-$$

Factors affecting the position of this equilibrium, and therefore the electrode potential, include the type of metal used, the concentration of metal ion solution, and the temperature of the system (Le Chatelier's principle – Chapter 7). The temperature and concentration are therefore standardized at 25 °C and 1 molar respectively. Under these conditions, the electrode potential of any particular half cell is known as the STANDARD ELECTRODE POTENTIAL, $E^{\ominus}$. In the case of a gas–gas ion half cell (see below) the gas pressure is standardized at 1 atmosphere pressure.

It is not possible to measure absolute single electrode potentials, since an electrical connection would have to be made between the solution and metal, and this would introduce another electrode potential into the system. However, it is possible to measure the potential difference between two connected half cells. If one of these half cells is a reference electrode, arbitrarily taken as having zero electrode potential, then potentials of all other electrodes can be measured against this reference. The reference electrode used is the hydrogen electrode (*Figure 9.1*) which consists of hydrogen gas at 1 atmosphere bubbling over an

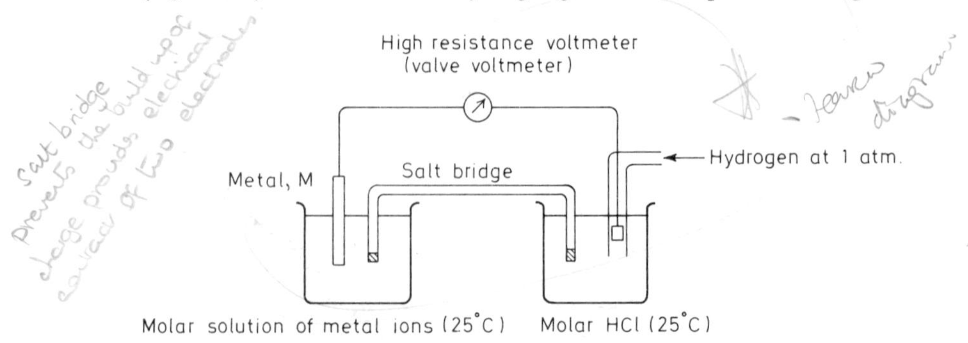

Figure 9.1 Measurement of standard electrode potential using the hydrogen electrode as reference

inert platinum surface which is dipped in a 1 molar solution of hydrogen ions. In practice, the hydrogen electrode is difficult to set up in the laboratory and so a secondary reference, the calomel electrode, is often used instead, which consists of mercury in contact with solid mercury(I) chloride and a solution of potassium chloride saturated with the mercury(I) salt.

If, in *Figure 9.1*, the half cell connected to the hydrogen electrode is zinc in 1 molar Zn^{2+} ion solution, it is found that the zinc electrode is the negative electrode, and that electrons flow from the zinc to the hydrogen electrode through the wire (and hence the hydrogen electrode is positive). A cell has been made, and the e.m.f. of the cell is measured as 0.76 V. Since the e.m.f. is the maximum potential difference of the cell, this value represents the $E^{\ominus}$ value for the $Zn^{2+}_{(aq)}|Zn_{(s)}$ half cell. The reason for this electron flow is that zinc, being more electropositive than hydrogen, preferentially ionizes to form Zn^{2+} ions in solution, leaving electrons on the zinc electrode. These electrons then flow through the connecting wire to the platinum surface of the hydrogen electrode where they are accepted by hydrogen ions to form hydrogen gas. These two electrode reactions can be represented by two HALF REACTIONS:

$$Zn \rightarrow Zn^{2+} + 2e^- \quad \text{and} \quad 2H^+ + 2e^- \rightarrow H_2$$

These two half reactions can now be combined to give the overall cell reaction:

$$Zn + 2H^+ \rightarrow Zn^{2+} + H_2$$

If, however, in *Figure 9.1*, the $Cu^{2+}_{(aq)}|Cu_{(s)}$ half cell is used instead of $Zn^{2+}_{(aq)}|Zn_{(s)}$, a voltage of 0.34 V is recorded. This time the copper electrode is found to form the positive of the cell which means that electrons flow from the hydrogen electrode to the copper, the half equations being:

$$H_2 \rightarrow 2H^+ + 2e^- \quad \text{and} \quad Cu^{2+} + 2e^- \rightarrow Cu$$

Since the anode is defined as the electrode at which positive electricity enters the solution (and *vice versa* for the cathode) then in the case of the $Zn^{2+}_{(aq)}|Zn_{(s)}$ half cell connected to the hydrogen electrode, the zinc electrode (−) is the anode and the hydrogen electrode (+) is the cathode. If the $Cu^{2+}_{(aq)}|Cu_{(s)}$ half cell is used instead of $Zn^{2+}_{(aq)}|Zn_{(s)}$ the hydrogen electrode (−) is the anode and the copper electrode (+) is the cathode: note that anode and cathode have opposite signs when electrolysis is considered.

In *Figure 9.1*, the SALT BRIDGE is a 'bridge' of electrolyte, which is often a concentrated solution of potassium chloride, or an agar gel made with concentrated potassium chloride solution. This serves two purposes: (1) provides electrical contact for the two electrolytes of the two half cells, and (2) prevents, by diffusion, build up of charge in the two beakers which, if allowed to occur, would eventually halt electron flow in the external circuit.

By considering other half cells connected to the hydrogen electrode, a table of standard electrode potentials can be compiled (*Table 9.1*). The sign of the $E^{\ominus}$ value for any particular half cell gives the sign of that electrode if connected to the hydrogen electrode. This table can now be used to calculate the e.m.f. developed by any combination of the above half cells. A familiar example is the DANIELL CELL, in which a $Zn^{2+}_{(aq)}|Zn_{(s)}$ and a $Cu^{2+}_{(aq)}|Cu_{(s)}$ half cell are connected together. This can be represented by the CELL DIAGRAM

$$Zn_{(s)}|Zn^{2+}_{(aq, 1M)}||Cu^{2+}_{(aq, 1M)}|Cu_{(s)}$$

Table 9.1 Cell reactions and their potentials

Half reaction	$E^{\ominus}_{298}$/V	*Half reaction*	$E^{\ominus}_{298}$/V
$K^+ + e^- \rightarrow K$	−2.92	$Cu^{2+} + 2e^- \rightarrow Cu$	+0.34
$Na^+ + e^- \rightarrow Na$	−2.71	$I_2 + 2e^- \rightarrow 2I^-$	+0.54
$Mg^{2+} + 2e^- \rightarrow Mg$	−2.37	$Ag^+ + e^- \rightarrow Ag$	+0.80
$Zn^{2+} + 2e^- \rightarrow Zn$	−0.76	$Br_2 + 2e^- \rightarrow 2Br^-$	+1.09
$Fe^{2+} + 2e^- \rightarrow Fe$	−0.44	$Cl_2 + 2e^- \rightarrow 2Cl^-$	+1.36
$2H^+ + 2e^- \rightarrow H_2$	0 (definition)	$F_2 + 2e^- \rightarrow 2F^-$	+2.87

where the single line represents the phase boundary and the double line the salt bridge. By convention, the more positive electrode is written on the right hand side of this diagram, and the total e.m.f. (at 25 °C) of the cell is given by

$$E_{cell} = E_{right} - E_{left}$$

Hence, from *Table 9.1*,

$$E_{cell} = 0.34 - (-0.76)\text{ V}$$
$$= 1.10\text{ V}$$

If, in fact, the above half cells are connected together under standard conditions, an e.m.f. of 1.10 V is recorded. The electrode reactions for this process are:

At zinc (−) Anode	At copper (+) Cathode
$Zn \rightarrow Zn^{2+} + 2e^-$	$Cu^{2+} + 2e^- \rightarrow Cu$

Overall reaction is $Zn + Cu^{2+} \rightarrow Zn^{2+} + Cu$

It should be remembered that these electrode reactions are reversible and, therefore, any concentration changes will affect the e.m.f. produced by the cell. If in the case of the Daniell cell, the Cu^{2+} ion concentration is increased, the cathode reaction is promoted further and so the e.m.f. will increase; on lowering the Cu^{2+} ion concentration, the cathode reaction is discouraged and the e.m.f. falls. Similarly, increasing the Zn^{2+} ion concentration decreases the e.m.f. and lowering the Zn^{2+} ion concentration increases the e.m.f. An equation (called the Nernst equation) does exist, which treats the effect of concentration on e.m.f. quantitatively (but is beyond the scope of this text).

From the above theory, elements can now be arranged in order of their standard electrode potentials, giving the ELECTROCHEMICAL SERIES, of which *Table 9.1* forms a part. Elements having the greatest negative $E^{\ominus}$ values are at the top of the series and are classed as the most electropositive. Uses of the series include:

(1) it gives the order of reactivity of the elements, i.e., metals decrease in reactivity, whilst non-metals increase in reactivity, as the series is descended – hence displacement reactions in solution can be predicted, e.g.,

$$Cu + 2Ag^+ \rightarrow Cu^{2+} + 2Ag$$

(2) it explains how iron is protected by coating with more reactive metals such as zinc. If the surface is scratched to expose the iron, rain water

(containing carbonic acid, an electrolyte) collects in the crack forming a simple cell, in which zinc is the anode and iron is the cathode; hence Zn^{2+} ions go into the solution, not Fe^{2+} ions (which would cause rusting);

(3) it can be used to predict products discharged at electrodes during electrolysis.

9.2 Oxidation and Reduction

Traditionally, oxidation was defined as the addition of oxygen or removal of hydrogen from a substance, whereas reduction was the removal of oxygen or the addition of hydrogen. The definition of oxidation has now been extended to include loss of electrons from an element, compound or ion whilst reduction is the gain of electrons by an element, compound or ion. Thus, if one substance loses electrons, another substance must gain them and so in a given reaction, if oxidation occurs, reduction must also occur; these reactions are called REDOX reactions. Therefore, all the cell reactions described above are redox reactions. In the Daniell cell, the overall cell reaction was established from two half equations, and this technique can also be used for balancing some of the more complicated redox reactions which occur (see below). In the Daniell cell it is clear that zinc is oxidized and the Cu^{2+} ion is reduced. However, in some other examples of redox reactions, the transfer of electrons is not always obvious and the OXIDATION STATE concept is therefore used to clarify this. The following is a set of rules for assigning oxidation states:

(1) Oxidation state of all elements in any allotropic form is zero.
(2) Oxidation state of oxygen is -2 in all its compounds except peroxides.
(3) Oxidation state of hydrogen is $+1$ in its compounds except when combined with a metal, where it is -1.
(4) All other oxidation states are selected in order to make the algebraic sum of the oxidation states equal to the net charge on the molecule or ion, e.g., in the MnO_4^- ion, manganese is therefore $+7$ so as to make the overall ionic charge (one) negative.
Consider the following reaction:

$$H_2S + Cl_2 \rightarrow 2HCl + S$$

Here, sulphur (in the H_2S) is oxidized since its oxidation state has been INCREASED from -2 to 0. Conversely, chlorine is reduced as its oxidation state is REDUCED from 0 to -1.

Consider the reaction between manganate(VII) ions and an iron(II) salt in acid solution, which can be written as

$$MnO_4^- + 8H^+ + 5Fe^{2+} \rightarrow 5Fe^{3+} + Mn^{2+} + 4H_2O$$

Note that manganese is reduced from oxidation state $+7$ to $+2$ whilst iron is oxidized from oxidation state $+2$ to $+3$. As previously mentioned, this overall redox equation can be constructed from the two half equations:

Oxidation: $Fe^{2+} \rightarrow Fe^{3+} + e^-$ (1)

Reduction: $MnO_4^- + 8H^+ + 5e^- \rightarrow Mn^{2+} + 4H_2O$ (2)

(using water to balance oxygens, then H^+ ions to balance the hydrogens, finally electrons to balance charge). Multiplication of equation (1) by 5 and addition to equation (2) to eliminate electrons from the chemical equation gives the overall reaction. Some common half reactions are given in *Table 9.2* for oxidizing and reducing agents. It should be emphasized that not all combinations of reducing

Table 9.2 Half reactions for oxidizing and reducing agents

Oxidizing agents

(1) $O_2 + 4e^- \rightarrow 2O^{2-}$ (O: $0 \rightarrow -2$)

(2) $Cr_2O_7^{2-} + 14H^+ + 6e^- \rightarrow 2Cr^{3+} + 7H_2O$ (Cr: $+6 \rightarrow +3$)

(3) $H_2O_2 + 2H^+ + 2e^- \rightarrow 2H_2O$ (O: $-1 \rightarrow -2$)

(4) $Cl_2 + 2e^- \rightarrow 2Cl^-$ (Cl: $0 \rightarrow -1$)

(5) $2H_2SO_4 + 2e^- \rightarrow SO_4^{2-} + 2H_2O + SO_2$ (S: $+6 \rightarrow +4$ in SO_2)

Reducing agents

(1) $H_2 \rightarrow 2H^+ + 2e^-$ (H: $0 \rightarrow +1$)

(2) $2I^- \rightarrow I_2 + 2e^-$ (I: $-1 \rightarrow 0$)

(3) $C_2O_4^{2-} \rightarrow 2CO_2 + 2e^-$

(4) $Sn^{2+} \rightarrow Sn^{4+} + 2e^-$ (Sn: $+2 \rightarrow +4$)

(5) $SO_3^{2-} + H_2O \rightarrow SO_4^{2-} + 2H^+ + 2e^-$ (S: $+4 \rightarrow +6$)

and oxidizing agents will lead to reaction, but the outcome of any particular reaction can often be predicted by consulting a table of $E^{\ominus}$ values, e.g.,

$$Cl_2 + 2I^- \rightarrow 2Cl^- + I_2 \quad (\textit{see Table 9.1})$$

9.2.1 EXTRACTION OF METALS FROM THEIR ORES

The method of extraction of a metal from its ore is governed by its electropositivity. The very electropositive metals K, Na, Mg and Al are very strongly bonded in their compounds and cannot be extracted by chemical reduction. In these cases, electrolytic reduction of the fused salt is employed (using the molten chlorides, except for Al in which the oxide is used – *see* p. 101). For metals of intermediate electropositivity (e.g., Zn, Fe) extraction is achieved by first obtaining the oxide (either naturally or by chemical means) and then converting this oxide into the metal by heating with carbon or carbon monoxide.

Example $2ZnS + 3O_2 \rightarrow 2ZnO + 2SO_2$

$$ZnO + C \rightarrow CO + Zn$$

Gold is so unreactive that it is found in its native state, and hence is merely refined.

Note that in some cases a metal can be obtained from its compound by heating with another metal. For example, titanium is found naturally as TiO_2; it is first converted into the tetrachloride and is then heated with either sodium or magnesium in an inert atmosphere of argon:

$$TiO_2 + 2C + 2Cl_2 \rightarrow TiCl_4 + 2CO$$

$$TiCl_4 + 4Na \rightarrow Ti + 4NaCl$$

10

KINETICS

10.1 Factors Affecting Reaction Rate

Chemical kinetics is the study of reaction rates and those factors which affect rate. This study not only helps in deciding the conditions to employ in reactions, but can also give information concerning the mechanism of a reaction. The main factors affecting reaction rate are (1) physical state of reactants, (2) concentration of reactants, (3) temperature, (4) the use of catalysts and (5) the use of light.

10.2 Physical State

Two solids mixed together rarely produce a reaction because movement (and therefore collisions) involving their constituent molecules, atoms or ions is restricted. If, however, the two reacting substances are mixed in solution, reaction can be immediate since the number of collisions, per unit time, which can now occur is greatly increased (because of the greater freedom of movement of particles involved).

In reactions between solids and liquids (or gases) reactions proceed more rapidly if the solid is powdered. This is because the effective surface area for reaction to occur has been increased; hence the number of collisions, per unit time, between reactants has also been increased.

10.3 Concentration

During a chemical reaction, the concentration of reactants decreases and hence the chance of collision between reactants is gradually reduced. This means that the rate of reaction decreases with time, and so a plot of reaction rate against time can take the form shown in *Figure 10.1(a)*. Rate of reaction can be measured in terms of rate of loss of reactants or rate of formation of products (the usual units of rate are mol l^{-1} s^{-1}).

Rate of reaction, therefore, is proportional to the concentration of reactants (*see* also Chapter 7). In the reaction,

$$2N_2O_5 \rightarrow 2N_2O_4 + O_2$$

it is found that

$$\text{Rate} \propto [N_2O_5] \quad \text{or} \quad \text{Rate} = k[N_2O_5]$$

The last equation is called the RATE EQUATION and k is called the RATE CONSTANT. Therefore, a plot of rate against $[N_2O_5]$ will give a straight line passing through the origin with a gradient of k [*see Figure 10.1(b)*]. Since $k = \text{rate}/[N_2O_5]$, then $k = \text{mol l}^{-1}\ \text{s}^{-1}/\text{mol l}^{-1}$, i.e., the units of k are s^{-1}.

However, in the reaction

$$2NO_2 \rightarrow 2NO + O_2$$

it is found that

$$\text{Rate} = k[NO_2]^2$$

Since $k = \text{rate}/[NO_2]^2$, the units of k are $\text{mol l}^{-1}\ \text{s}^{-1}/\text{mol}^2\ \text{l}^{-2} = \text{mol}^{-1}\ \text{l s}^{-1}$. Note that in the rate equation for the decomposition of dinitrogen pentoxide, the $[N_2O_5]$ term is raised to the power of one, and is therefore said to be a FIRST ORDER reaction, whilst in the rate equation for the decomposition of nitrogen dioxide, the $[NO_2]$ term is raised to the power of two, and is said to be a SECOND ORDER reaction. Therefore, in the general reaction:

$$aA + bB \rightarrow \text{products}$$

if the rate equation has the form, rate $= k[A]^x[B]^y$, then the reaction is said to be of order x with respect to A, y with respect to B, and the OVERALL ORDER of the reaction is $x + y$, i.e., overall order is the sum of the powers to which the concentrations are raised in the rate equation. It must be emphasized that a rate law, and therefore order, of a given reaction must be determined experimentally, and cannot be predicted from the chemical equation (the decomposition of N_2O_5 and NO_2 illustrate this point).

Most chemical reactions proceed by a sequence of distinct steps, each step having a different rate. The overall rate of the reaction is controlled or dominated by the slowest step, which is termed the RATE-DETERMINING STEP. The number of atoms or molecules taking part in each step leading to chemical reaction is known as the MOLECULARITY, and although molecularity and order are often identical, they are not always the same.

There are several ways of measuring the rate of reaction.

(*a*) Gas measurements: if one of the products of reaction is a gas, the volume of gas evolved with time can be measured.

(*b*) Titration: if one of the products is an acid, e.g., hydrolysis of an ester, then samples of the reaction mixture can be removed (and quenched if necessary to stop the reaction) at various times, and the amount of acid formed is determined by titration.

(*c*) Conductivity measurements: if a rapidly moving ion is either removed or produced during a reaction, then the reaction can be followed by measuring the conductance of the solution with time.

(*d*) Colorimetry: if a reactant or product is coloured, then the colour of the reaction mixture will either diminish or become more intense as the reaction proceeds. By directing light of wavelengths absorbed by the solution onto the reaction flask, and allowing the emerging light to then fall on a light-sensitive cell (which gives an e.m.f. proportional to the light intensity), the reaction can be followed with time.

Suppose the rate of the reaction $A \rightarrow B + C$ is followed by one of the above methods. If a graph of [A] against time is plotted [*Figure 10.1(a)*], then the rates of reaction at various concentrations can be determined from the slopes

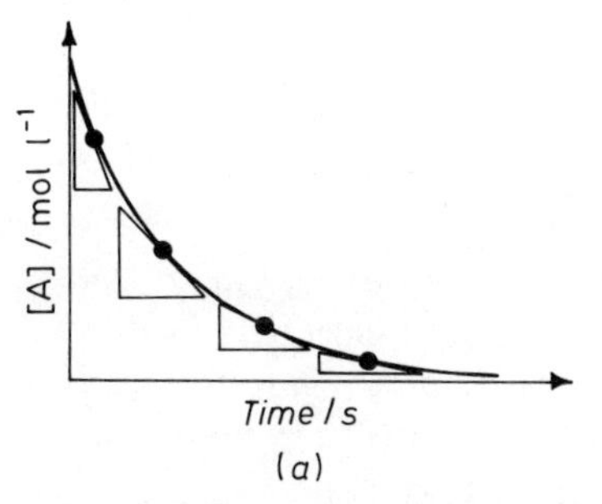

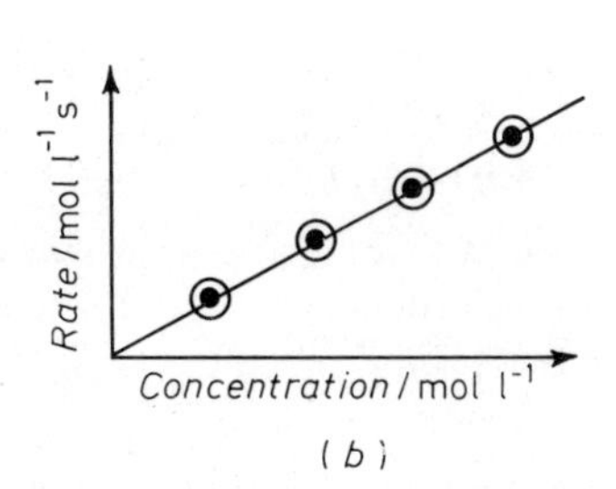

Figure 10.1 Plots of (a) concentration against time and (b) rate against concentration for the first order reaction A → B + C

of the tangents drawn to the curve at those concentrations. The reaction is first order if a plot of rate against concentration is a straight line [rate = k[A]; *Figure 10.1*(*b*)], but second order, for example, if a plot of rate against (concentration)^2 is a straight line (rate = $k[A]^2$).

It can be seen from *Figure 10.1*(*a*) that the time taken for the concentration of A to fall to one half its initial value is the same as the time taken for the concentration to fall from one half to one quarter its initial value, etc. The half-life, $t_{1/2}$, for a first order reaction, therefore, is constant. Radioactive decay is an example of a first order process (*see* Chapter 1).

10.4 Temperature

It is found that a small increase in the temperature of a reaction results in a relatively large increase in reaction rate. When the temperature is raised, (*a*) the collision rate of reactant molecules is increased and (*b*) the average energy of reactant molecules is increased, and hence a greater proportion of these molecules has the required minimum energy for reaction to occur (it is this latter increase which contributes mainly to the increased rate). This minimum energy is called the ACTIVATION ENERGY, E_a, for the reaction. At some stage during reaction, the molecules are rearranging themselves to form products. At this stage, therefore, they must be in an unstable or high energy state, resulting in an energy barrier between reactants and products. The height of this barrier represents E_a, and reacting molecules, when at the top of this barrier, are said to form an ACTIVATED COMPLEX or be in a TRANSITION STATE; *Figures 10.2*(*a*) and (*b*) illustrate this for both exothermic and endothermic reactions (*see* also Chapter 5). Consequently, reacting molecules must possess sufficient energy to surmount this barrier.

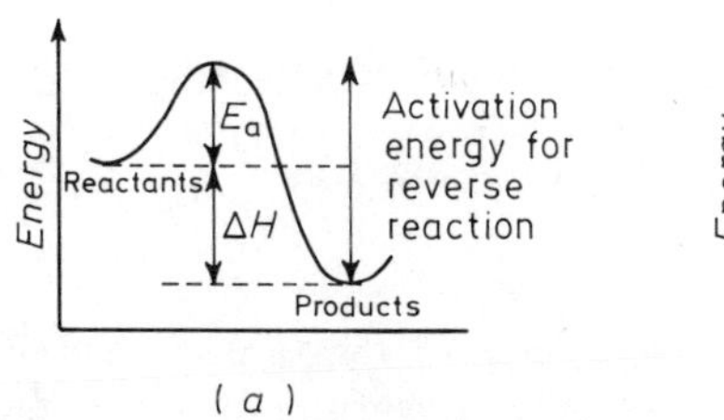

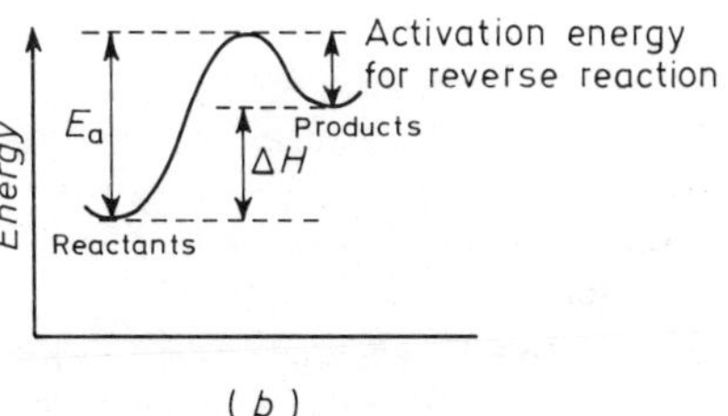

Figure 10.2 Energy profiles during (a) an exothermic reaction and (b) an endothermic reaction

The relationship between rate constant, temperature and activation energy is given by the ARRHENIUS EQUATION

$$k = Ae^{-E_a/RT} \tag{10.1}$$

where k = rate constant, E_a = activation energy, R is the gas constant, T is the absolute temperature and A is a constant. Note that the exponential function describes the large increase in rate for relatively small increases in temperature.

Figure 10.3 (a) Effective and (b) ineffective collisions in the reaction $2HI \rightarrow H_2 + I_2$

It should be noted that even for collisions (between reactants) with energy in excess of E_a, reaction may not occur if colliding molecules are not correctly aligned, i.e., there are STERIC requirements for reaction, and this is illustrated in *Figure 10.3* for the reaction $2HI \rightarrow H_2 + I_2$ [this steric requirement is incorporated into constant A in equation (10.1)].

10.5 Catalysts

A catalyst is a substance which alters the rate of a reaction without itself undergoing any permanent chemical change. Catalysts function by providing an alternative route for the reaction. If this alternative route has a lower activation energy than the uncatalysed reaction, the catalyst is a POSITIVE CATALYST, and reaction rate increases. A substance which can lower the rate of a reaction, however, is said to be a NEGATIVE CATALYST or INHIBITOR. A catalyst does not alter the position of equilibrium; it only alters the rate at which equilibrium is attained.

A HOMOGENEOUS CATALYST is one which is in the same phase as the reactants, e.g., the acid-catalysed production of an ester (*see* Chapter 15, p. 177), whereas a HETEROGENEOUS CATALYST is one which is in a different phase from the reactants, e.g., decomposition of hydrogen peroxide (liquid) using manganese(IV) oxide (solid), or the Contact and Haber processes (*see* Chapter 7). Two main theories of catalysis exist:

(*a*) The intermediate compound theory, which applies mainly to homogeneous catalysis. In this situation, the catalyst is continuously consumed and regenerated; an example of this is the halogenation of benzene using aluminium halide catalyst (*see* Chapter 15, p. 161).

(*b*) The adsorption theory, which applies to heterogeneous catalysis. Here, reactant molecules collide with, and are adsorbed onto, the catalyst surface; reaction occurs and the products are then desorbed leaving the surface free for adsorption of more reactants. An example of this type is the hydrogenation of alkenes using a nickel catalyst (*see* Chapter 15, p. 155).

10.6 Light

Under the influence of ultraviolet light, methane reacts with chlorine to give chlorinated compounds. The ultraviolet light serves to split the chlorine molecules into neutral atoms each having an unpaired electron (called FREE RADICALS). These chlorine radicals can now ABSTRACT hydrogen atoms as shown

$$Cl_2 \xrightarrow{uv} Cl\cdot + Cl\cdot$$

Then

$$CH_4 + Cl\cdot \rightarrow \underset{\text{methyl radical}}{CH_3\cdot} + HCl$$

$$CH_3\cdot + Cl_2 \rightarrow \underset{\text{chloromethane}}{CH_3Cl} + Cl\cdot$$

$$CH_3Cl + Cl\cdot \rightarrow \underset{\text{chloromethyl radical}}{CH_2Cl\cdot} + HCl$$

$$CH_2Cl\cdot + Cl_2 \rightarrow \underset{\text{dichloromethane}}{CH_2Cl_2} + Cl\cdot$$

$$CH_2Cl_2 + Cl\cdot \rightarrow \underset{\text{dichloromethyl radical}}{CHCl_2\cdot} + HCl$$

$$CHCl_2\cdot + Cl_2 \rightarrow \underset{\text{trichloromethane}}{CHCl_3} + Cl\cdot$$

$$CHCl_3 + Cl\cdot \rightarrow \underset{\text{trichloromethyl radical}}{CCl_3\cdot} + HCl$$

$$CCl_3\cdot + Cl_2 \rightarrow \underset{\text{tetrachloromethane}}{CCl_4} + Cl\cdot$$

Figure 10.4 Free radical substitution of methane by chlorine

(*Figure 10.4*), to form hydrogen chloride and new radicals (e.g., $CH_3\cdot$, $CH_2Cl\cdot$ etc): Such reactions are called PHOTOCHEMICAL reactions.

10.7 Reaction Mechanisms

Kinetics data can elucidate reaction mechanisms. For example, in the hydrolysis of bromomethane in aqueous alkali, the rate equation is given by rate $= k[CH_3Br][OH^-]$ and the mechanism is thought to be

$$OH^- + CH_3Br \rightarrow \underset{\text{(Transition state)}}{[HO\text{------}CH_3\text{------}Br]^-} \rightarrow CH_3OH + Br^-$$

where the dotted lines represent partial bonds. The reaction is bimolecular and second order.

However, the hydrolysis of 2-chloro-2-methylpropane (t-butyl chloride) follows a rate equation:

$$\text{Rate} = k[(CH_3)_3CCl]$$

and so the rate is independent of $[OH^-]$. The mechanism is thought to be

$$(CH_3)_3CCl \xrightarrow[\text{(rate-determining)}]{\text{slow}} (CH_3)_3C^+ + Cl^- \xrightarrow[\text{fast}]{OH^-} (CH_3)_3COH$$

and so only $(CH_3)_3CCl$ is involved in the rate-determining step. The hydrolysis is therefore a unimolecular first order reaction.

11

PERIODICITY

11.1 Trends in the Periodic Table

When a period is crossed from left to right, there is a regular increase (from one to eight) in the number of outer shell electrons. Hence, trends in periods (and groups) can be observed, and can be explained in terms of atomic structure (*see* also Introduction, p. 6). Some of these trends are now discussed.

11.2 Ionization Energy and Electron Affinity

The ionization energy of an element (defined on p. 3) can be determined from its atomic emission spectrum, since the convergence limit (*see Figure 3*) represents the point at which an electron becomes detached from the atom. Use of the equation $E = h\nu$ (p. 5) can then give the value of the ionization energy. Alternatively, it can be determined by an electron bombardment method; this involves the use of a radio valve (thyratron) containing gas at low pressure. Here, the cathode is heated by an electric current, and electrons are emitted. These can be accelerated towards a positive grid and, on passing through the grid, are then repelled by making the anode of the valve negative; hence no current flows. If, however, the electrons are travelling fast enough (their velocities are controlled by the magnitude of the potential difference between cathode and grid), they will ionize the gas atoms (or molecules) and current then flows because the resulting positive ions are attracted to the anode. The ionization energy can then be found from this potential difference (between cathode and grid) at which ionization of the gas occurs.

Ionization energies decrease as a group is descended (the reasons for this are outlined on p. 7) and, generally speaking, increase as a period is crossed from left to right (p. 7), as shown in *Figure 2*. The minor irregularities in the I_1 values across, for example, the third period (Na to Ar) indicate the relative stabilities of completely filled or half-filled sub-shells.

The electron affinity of an element (defined on p. 7) can be determined from a Born–Haber cycle if the lattice energy of a compound containing the element is estimated by calculation (*see* Chapter 5). The general trend is that electron affinity increases across any particular period because of the increasing nuclear charge acting on the same shell (*see* p. 7).

11.3 Atomic and Ionic Radii

The size of an atom can be expressed in terms of its covalent radius. This is half the internuclear distance in a diatomic molecule of the element (in the case of, say, lithium this refers to lithium vapour, Li_2). If the covalent radius, therefore, is plotted against atomic number (*Figure 4*), it is seen that atomic radius decreases as a period is crossed from left to right, and increases on descending a group (the reasons for this are discussed on p. 6).

The size of an ion varies somewhat according to its environment, and, therefore, average values are used. *Figure 4* shows a plot of ionic radius against atomic number, and the reasons for these values are outlined on p. 8).

11.4 Atomic Volume, Melting Point, Boiling Point and Latent Heats of Fusion and Vaporization

Periodicity is shown by physical properties such as atomic volume (volume/cm^3 of 1 mole of the solid element – *see Figure 11.1*), melting point, boiling point and the latent heats (or enthalpies) of fusion and vaporization. Some of these values, for H to Kr, are given in *Table 11.1*. Overall trends are not always clear for these properties since they do not concern just isolated atoms – chemical structure also has to be considered, i.e., some elements exist as atoms, some as diatomic molecules, some as macromolecules, some as metallic lattices, and so on. For example, data given in *Table 11.1* (which should be studied in conjunction with Chapters 12 and 13) indicate the high melting point of carbon compared with nitrogen; this is because carbon exists as a macromolecule whereas nitrogen exists as diatomic molecules. Also illustrated are the low melting points of the Group I alkali metals compared with the transition metals, which would imply that in the latter case, 3*d* and 4*s* electrons are delocalized in the metallic lattice, resulting in greater stabilization and stronger metallic bonds (this is also

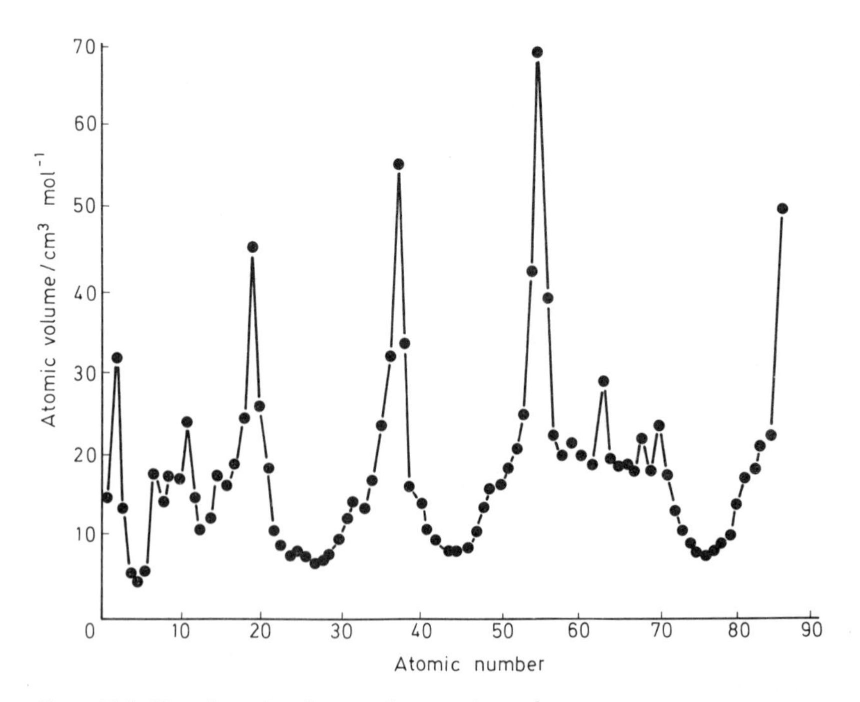

Figure 11.1 Plot of atomic volume against atomic number

Table 11.1 Melting points, boiling points, and enthalpies of fusion of elements

	H	He
M.p./K	14	4*
B.p./K	20	4
$\Delta H^{\ominus}_{m}$	0.06	0.02*

	Li	Be	B	C_d	N	O	F	Ne
M.p./K	454	1556	2300	3823	63	54	53	25
B.p./K	1604	2750	4200	5100	77	90	85	27
$\Delta H^{\ominus}_{m}$	3.01	11.72	22.18	–	0.36	0.22	0.26	0.33

	Na	Mg	Al	Si	P_w	S_m	Cl	Ar
M.p./K	371	923	932	1683	317	392	172	84
B.p./K	1163	1390	2720	2950	554	718	239	87
$\Delta H^{\ominus}_{m}$	2.60	8.95	10.75	46.44	0.63	1.41	3.20	1.18

	K	Ca	Sc	Ti	V	Cr	Mn	Fe	Co	Ni	Cu	Zn	Ga	Ge	As	Se	Br	Kr
M.p./K	336	1123	1673	1950	2190	2176	1517	1812	1768	1728	1356	693	303	1210	1090*	490	266	116
B.p./K	1039	1765	2750	3550	3650	2915	2314	3160	3150	3110	2855	1181	2510	3100	886_s	958	331	120
$\Delta H^{\ominus}_{m}$	2.30	8.66	16.11	15.48	17.57	13.81	14.64	15.36	15.23	17.61	13.05	7.38	5.59	31.80	27.61*	5.23	5.29	1.64

$\Delta H^{\ominus}_{m}$ = enthalpy of fusion in kJ mol^{-1}; symbols of elements followed by a subscript denote allotropes; * denotes measurements made under pressure; s denotes sublimation at that temperature.

shown by the higher enthalpies of fusion). Other trends which can be seen include the increasing melting and boiling points as Group 0 is descended (because of increasing van der Waals forces), and the decreasing melting and boiling points as Group I is descended (because of the weaker metallic bonds). It should also be noted that metals occupy mainly the left and centre of the Periodic Table whilst the non-metals are mainly on the right hand side of the table.

11.5 Chemical Properties and Periodicity

As explained in the Introduction, there is a decrease in electropositivity (or an increase in electronegativity) across the Periodic Table from left to right, and an increase in electropositivity down a group. *Tables 11.2, 11.3* and *11.4* show some of the effects which the former trend has on the properties of the oxides, chlorides and hydrides of the third period elements.

Table 11.2 Oxides

Formula	Na_2O Na_2O_2	MgO	Al_2O_3	SiO_2	P_4O_6 P_4O_{10}	SO_2 SO_3	Cl_2O
Physical state (at 293 K)	Solids	Solid	Solid	Solid	Solids	Gas Liquid	Gas
Bonding	Ionic	Ionic	Ionic	Covalent	Covalent	Covalent	Covalent
Classification	Basic	Basic	Amphoteric	Acidic	Acidic	Acidic	Acidic

Table 11.3 Chlorides

Formula	$NaCl$	$MgCl_2$	$AlCl_3$	$SiCl_4$	PCl_3 PCl_5	S_2Cl_2
Physical state (at 293 K)	Solid	Solid	Solid	Liquid	Liquid Solid	Liquid
Bonding	Ionic	Ionic	Covalent	Covalent	Covalent Ionic	Covalent
Action of moist air	No effect	No effect	Fumes	Fumes	Fumes	Fumes

Table 11.4 Hydrides

Formula	NaH	MgH_2	$(AlH_3)_n$	SiH_4	PH_3	H_2S	HCl
Physical state (at 293 K)	Solid	Solid	Solid	Gas	Gas	Gas	Gas
Bonding	Ionic	Ionic	Covalent	Covalent	Covalent	Covalent	Covalent

11.5.1 DIAGONAL RELATIONSHIPS

Because electropositivity increases down a group and decreases across a period (from left to right), elements diagonally below one another have similar electropositivities (and therefore properties), as shown below:

Li ⟷ Mg, Be ⟷ Al, B ⟷ Si

Li	Be	B	C
Na	Mg	Al	Si

Hence, both Li^+ and Mg^{2+} ions become hydrated because of their small size; lithium and magnesium carbonates have low solubilities in water, and both decompose on heating to evolve carbon dioxide; both lithium and magnesium hydroxides are weakly basic in water; both lithium and magnesium form nitrides.

Similarly, beryllium and aluminium have similar properties; both show mainly covalent character in their compounds; both metals are resistant to acids, unless finely divided, because of the impervious oxide layers which form on their surfaces; both metals dissolve in strong bases to give beryllates and aluminates; both beryllium and aluminium oxides are amphoteric.

Similarly, there are common properties for boron and silicon (and some of their compounds).

12

SOME ELEMENTS AND GROUPS OF THE PERIODIC TABLE

12.1 Hydrogen

Hydrogen is a colourless, odourless and tasteless gas; it is also the least dense gas known. Free hydrogen occurs to probably less than one part in a million in the atmosphere. Some physical properties of hydrogen are given in *Table 12.1.*

Table 12.1 Hydrogen

Electronic configuration	$1s^1$	*Enthalpy of atomization*/kJ mol^{-1}	218
Ionization energy, I_1 /kJ mol^{-1}	1310	*Atomic (covalent) radius*/nm	0.037
Electron affinity/kJ mol^{-1}	−72	*Hydration energy of* H^+ *ion*/ kJ mol^{-1}	−1075

Hydrogen shows similarities to the Group I alkali metals. For example, hydrogen has one electron in its outer shell and so forms a positive (H^+) ion; however, the electron is difficult to remove and so its I_1 value (*Table 12.1*) is high (compare with Li and Na, which have I_1 values of 520 and 500 kJ mol^{-1} respectively). Hydrogen also shows similarities to the Group VII halogens since (1) its molecules are diatomic and (2) it forms H^- ions to attain noble gas structure. However the latter process occurs less readily than for the halogens (the electron affinities for fluorine and chlorine are −332.6 and −364 kJ mol^{-1} respectively; cf. value for hydrogen, −72 kJ mol^{-1}). Also shown in *Table 12.1* is the high value for the enthalpy (heat) of atomization of hydrogen (the values for fluorine and chlorine are 79.1 and 121.1 kJ mol^{-1} respectively), i.e., the H–H bond is relatively strong, since the shared electron pair 'cloud' attracts the hydrogen nuclei to it (and hence to each other) strongly (there are no non-bonding electrons to screen this attraction) – *see* Introduction, p. 13.

Because of the proton's small size, it has a unique ability to distort the electron 'cloud' surrounding other atoms, and consequently the proton never exists on its own except in gaseous ion beams. In aqueous solution the H^+ ion is always hydrated; the highly exothermic value of its hydration energy is given in *Table 12.1.*

12.1.1 PREPARATIONS OF HYDROGEN

Several laboratory preparations exist, some of which are listed below.

12.1.1.1 From Acids

The action of a metal above hydrogen in the electrochemical series with a non-oxidizing acid will cause the reduction of protons to hydrogen gas.

$$2H^+ + Zn \rightarrow Zn^{2+} + H_2$$

However, nitric acid, being a stronger oxidizing agent than the proton, is therefore preferentially reduced if used, to produce oxides of nitrogen in addition to, or instead of, hydrogen.

The electrolysis of, say, dilute sulphuric acid will also produce hydrogen by the cathode reaction:

$$2H^+ + 2e^- \rightarrow H_2$$

12.1.1.2 From Water

The more electropositive metals are able to release hydrogen from water or steam:

$$Ca + 2H_2O_{(l)} \rightarrow Ca(OH)_2 + H_2$$

$$3Fe + 4H_2O_{(g)} \rightleftharpoons Fe_3O_4 + 4H_2$$

12.1.1.3 From Alkalis

Silicon and metals such as aluminium and tin produce hydrogen if added to solutions of strong alkalis:

$$2Al + 2OH^- + 6H_2O \rightarrow 2[Al(OH)_4]^- + 3H_2$$

12.1.1.4 From Ionic Hydrides

The addition of water to an ionic hydride releases hydrogen.

$$NaH + H_2O \rightarrow NaOH + H_2$$

12.1.2 HYDRIDES

Ionic hydrides are produced when strongly electropositive metals are heated with hydrogen at temperatures up to 700 °C, to form white crystalline solids:

$$2Na + H_2 \rightarrow 2NaH$$

$$Ca + H_2 \rightarrow CaH_2$$

These hydrides are hydrolysed by water (*see* above).

Covalent hydrides are formed by the combination of most non-metals with hydrogen. Generally speaking, covalent hydrides are gaseous at room temperature. It would be expected that the hydrides of the elements of any particular group would have gradually increasing boiling points as the group is descended. This is generally true, but anomalies occur because of hydrogen bonding (*see* p. 25 and *Table 3.1*). This is particularly evident for H_2O and H_2S (hydrides of Group VI elements).

Hydrides are also known where the hydrogen is present in a complex anion, e.g., lithium tetrahydridoaluminate (lithium aluminium hydride), $LiAlH_4$

(prepared as below), and sodium tetrahydridoborate (sodium borohydride), $NaBH_4$.

$$4LiH + AlCl_3 \xrightarrow{\text{ether}} LiAlH_4 + 3LiCl$$

Both of these hydrides are strong reducing agents, and are particularly useful in organic chemistry for reducing carboxylic acids and aldehydes to primary alcohols, and ketones to secondary alcohols (*see* Chapter 15).

12.2 Groups I and II (Li–Cs and Be–Ba)

Elements in Groups I and II have one and two electrons respectively in their outer shell, and so there is a tendency for them to lose the outer electron(s) to form positive ions. *Table 12.2* gives some of the physical properties of the Group I and II elements.

Table 12.2 Metals of Groups I and II

Element	*Electronic configuration*	I_1 / kJ mol^{-1}	*Atomic radius* / nm	*Ionic radius* / nm	*Element*	*Electronic configuration*	$I_1 + I_2$ / kJ mol^{-1}	*Atomic radius* / nm	*Ionic radius* / nm
Li	[He]$2s^1$	520	0.123	0.068	Be	[He]$2s^2$	2700	0.106	0.030
Na	[Ne]$3s^1$	500	0.157	0.098	Mg	[Ne]$3s^2$	2240	0.140	0.065
K	[Ar]$4s^1$	420	0.203	0.133	Ca	[Ar]$4s^2$	1690	0.174	0.094
Rb	[Kr]$5s^1$	400	0.216	0.148	Sr	[Kr]$5s^2$	1650	0.191	0.110
Cs	[Xe]$6s^1$	380	0.235	0.167	Ba	[Xe]$6s^2$	1500	0.198	0.134

Trends down the groups include:

(1) Ionization energies decrease (because outer electrons are further from the nucleus, and more efficient electron screening or shielding occurs) and so electropositive character increases and metals become more reactive.
(2) Ionic size (*Table 12.2*) increases and hence: hydroxides become stronger bases (since charge density decreases and so there is less attraction between the M^+ or M^{2+} ion and the OH^- ion); salts have a decreasing tendency to become hydrated (because the decreasing charge density of the metal ion means a smaller attraction for the $\delta-$ charge on the oxygen atoms of water molecules); lattice energies of their compounds decrease (because of decrease in charge density) (*see Table 12.3*), but note that corresponding lattice energies for Group II elements are greater because the cations are doubly charged (*see* also *Table 12.3*); carbonates are more stable to heat (*see* later).

12.2.1 ACTION OF THE ELEMENTS WITH WATER, CHLORINE AND OXYGEN

12.2.1.1 Water

All the Group I metals readily react with water to form the hydroxide and hydrogen, the rate of reaction increasing as the group is descended (lithium is comparatively slow whereas caesium is violent); these reactions can be represented as

$$2M + 2H_2O \rightarrow 2MOH + H_2$$

Of the Group II metals, beryllium does not react with water, even when hot, magnesium reacts very slowly with water but readily with steam at red heat:

$$Mg + H_2O \rightarrow MgO + H_2$$

and calcium, strontium and barium all react readily with water, e.g.,

$$Ca + 2H_2O \rightarrow Ca(OH)_2 + H_2$$

12.2.1.2 Chlorine

The elements all react with chlorine to form ionic chlorides (except beryllium which forms a solid chloride having poor electrical conductivity in the molten state, hence indicating appreciable covalent character). The lattice energies of the chlorides are found to decrease as the cation size increases (*Table 12.3*) because of the decrease in charge density of the cation.

12.2.1.3 Oxygen

All elements react with oxygen but many of the Group I metals are able to form peroxides and superoxides (K, Rb, Cs give yellow to orange crystalline solids called superoxides of formula MO_2); the latter are not formed by the Group II elements.

$$Li \xrightarrow{O_2} Li_2O$$

$$Na \xrightarrow{O_2} Na_2O \xrightarrow{O_2} Na_2O_2$$

$$K \xrightarrow{O_2} K_2O \xrightarrow{O_2} K_2O_2 \xrightarrow{O_2} KO_2$$

(same for Rb and Cs)

$$Be \xrightarrow{O_2} BeO$$

$$Mg \xrightarrow{O_2} MgO$$

$$Ca \xrightarrow{O_2} CaO$$

$$Sr \xrightarrow{O_2} SrO \xrightarrow{O_2} SrO_2$$

(same for Ba)

12.2.2 SOLUBILITY OF SALTS

The solubility of a salt can depend on two opposing factors; the lattice energy and the hydration energy (*see* Chapter 5). Low solubility can result if lattice energy exceeds hydration energy (although entropy factors have to be considered). Group I salts tend to be readily soluble in water (although some lithium salts, e.g., LiOH, Li_2CO_3, LiF, exhibit low solubilities because of the small size of the Li^+ ion, giving rise to high lattice energies) whereas many of those formed by Group II elements (particularly if the salts contain a bivalent anion, e.g., SO_4^{2-}, CO_3^{2-}, $C_2O_4^{2-}$) have limited solubility because of the increased lattice energy. In the case of the sulphates of Group II, there is a marked decrease in solubility as cation size increases ($MgSO_4$ soluble, $CaSO_4$ sparingly soluble, $BaSO_4$ insoluble). This is explained by a corresponding decrease in hydration energy.

Table 12.3 Lattice energies of chlorides and hydration energy of cations

Element	*Lattice energy of chloride/* kJ mol^{-1}	*Hydration energy of cation/* kJ mol^{-1}	*Element*	*Lattice energy of chloride/* kJ mol^{-1}	*Hydration energy of cation/* kJ mol^{-1}
Li	−849	−499	Be	−2983	–
Na	−781	−390	Mg	−2489	−1891
K	−710	−305	Ca	−2197	−1562
Rb	−685	−281	Sr	−2109	−1414
Cs	−641	−248	Ba	−1958	−1273

12.2.3 PROPERTIES OF COMPOUNDS

12.2.3.1 Oxides

Those of Group I dissolve readily in water to form strongly alkaline solutions (except Li_2O, which does not form a strongly alkaline solution), whilst those of Group II are of limited solubility, e.g.,

$$Na_2O + H_2O \rightarrow 2NaOH$$

$$CaO + H_2O \rightarrow Ca(OH)_2$$

In both groups, the base strengths of the resulting solutions increase down the group (*see* before).

12.2.3.2 Hydroxides

Group I metal hydroxides are thermally stable (except LiOH) whilst those of Group II decompose to give the oxide, e.g.,

$$Ca(OH)_2 \rightarrow CaO + H_2O$$

Sodium and potassium hydroxides are used extensively both in preparative work and quantitative work, such as volumetric analysis.

12.2.3.3 Nitrates

On heating, Group I nitrates decompose to give the nitrite (except $LiNO_3$, which gives the oxide) whilst Group II nitrates form the oxide, e.g.,

$$2NaNO_3 \rightarrow 2NaNO_2 + O_2$$

$$2Ca(NO_3)_2 \rightarrow 2CaO + 4NO_2 + O_2$$

12.2.3.4 Carbonates

The Group I carbonates dissolve in water (Li_2CO_3 has limited solubility) to give alkaline solutions (*see* salt hydrolysis: Chapter 8) and are stable to heat, except Li_2CO_3 which decomposes on strong heating:

$$Li_2CO_3 \rightarrow Li_2O + CO_2$$

This is because there is a large gain in electrostatic attraction between the very small cation, Li^+, and the smaller O^{2-} ion instead of the much larger CO_3^{2-} ion (this gain is relatively much less for the larger Group I cations, so the other Group I carbonates are thermally stable).

In contrast, Group II carbonates are insoluble and decompose on heating to give the oxide and carbon dioxide.

12.2.3.5 Hydrogencarbonates (bicarbonates)

These are formed by passing carbon dioxide into a solution or suspension of the carbonate, e.g.,

$$CaCO_3 + H_2O + CO_2 \rightarrow Ca(HCO_3)_2$$

Only those of Group I can be isolated in the solid state, but all decompose on heating, e.g.,

$$2KHCO_3 \rightarrow K_2CO_3 + H_2O + CO_2$$

$$Ca(HCO_3)_2 \rightarrow CaCO_3 + H_2O + CO_2$$

12.2.3.6 Nitrides

Group II metals and lithium form nitrides by heating the elements in nitrogen, e.g.,

$$3Mg + N_2 \rightarrow Mg_3N_2$$

12.2.3.7 Halides

The halides, many of which are prepared by the action of the halogen acid on the metal, oxide, hydroxide or carbonate, are not hydrolysed in solution but can be if the polarizing power of the cation is high, i.e., LiCl is slightly hydrolysed and $BeCl_2$ is readily hydrolysed when their aqueous solutions are heated. Lithium and beryllium chlorides are both soluble in some organic solvents, showing that they both exhibit some covalent character.

12.2.4 DIFFERENCES OF THE HEAD ELEMENTS OF THE GROUPS

The above properties show how the head elements of a group can show abnormal properties, explained, in many cases, by the small atomic and ionic sizes of the elements involved. Lithium, for example, forms some compounds which have low solubilities, forms a nitride directly, and its hydroxide, carbonate and nitrate decompose on heating to form the oxide; these properties are similar to those of magnesium (*see* diagonal relationships, p. 90). Beryllium and lithium both form some compounds which are covalent or partially covalent (*see* Fajans's rules, p. 13).

12.3 Group III (Boron and Aluminium)

Boron is a non-metal whilst aluminium is a metal. Unlike the Group I and II metals, boron and aluminium do not readily lose their (three) outer shell electrons to attain noble gas configurations. The ionization energy for this process is so high in the case of boron (*see Table 12.4*) that it never forms the B^{3+} ion under ordinary circumstances. Even for aluminium, where this energy is slightly less (*Table 12.4*), the simple Al^{3+} ion is only found in the fluoride and oxide (and even here the bonds may be partly covalent, since the Al^{3+} ion is small (*Table 12.4*) and highly polarizing – Fajans's rules). However, the hydrated ion $[Al(H_2O)_6]^{3+}$ is known in aqueous solution (since the hydration energy compensates for the high ionization energy), and in solid hydrated salts, e.g., $Al_2(SO_4)_3.16H_2O$. Aluminium, therefore, generally forms covalent bonds.

Table 12.4 Boron and aluminium

Element	*Electronic configuration*	*Ionization energy* $I_1+I_2+I_3$/kJ mol^{-1}	*Atomic radius*/nm	*Ionic radius*/nm
B	[He] $2s^2 2p^1$	6900	0.080	0.016
Al	[Ne] $3s^2 3p^1$	5080	0.125	0.045

The main group valency is 3, since one of the outer *s* electrons is promoted to a *p* orbital to give three unpaired electrons available for bonding. Consequently, noble gas structures are not achieved by B or Al when their covalencies are 3, e.g., in BF_3, but noble gas structure (or the octet) can be achieved if such molecules accept an electron pair from another source, e.g., $H_3N\!: \rightarrow AlCl_3$. Here, the $AlCl_3$ is acting as a Lewis acid (p. 65); $AlCl_3$ makes use of this property to generate electrophilic sites in organic preparative work (*see* p. 161). Note that the covalency for boron is a maximum of 4, e.g., as in the tetrafluoroborate ion, $[BF_4]^-$, since the second shell can only hold eight electrons. For aluminium, however, it is 6, e.g., $[AlF_6]^{3-}$ as in Na_3AlF_6, sodium hexafluoroaluminate (cryolite), since the *d* orbitals in aluminium's third shell can be used: note that although there are three unoccupied *d* orbitals remaining, no more than six F^- ions can fit round the small Al^{3+} ion.

12.3.1 The Halides

All the trihalides of boron are covalent, their shapes being trigonal planar, bond angles 120 degrees (*see* Chapter 2). At room temperature, BF_3 and BCl_3 are gases, BBr_3 is a liquid and BI_3 is a solid. They are attacked by water because the boron is electron deficient (*see* before); hydrolysis gives the hydrogen halide, except for boron trifluoride which gives HBF_4:

$$BCl_3 + 3H_2O \rightarrow H_3BO_3 + 3HCl$$

$$4BF_3 + 3H_2O \rightarrow 3HBF_4 + H_3BO_3$$

The hydrolysis reactions are thought to proceed as in *Figure 12.1.*

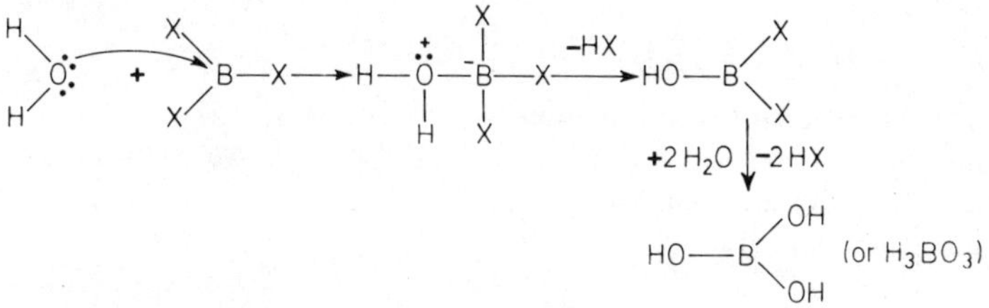

Figure 12.1

Of the aluminium halides, aluminium fluoride is ionic, whilst the rest are covalent. The fluoride can be prepared by heating solid ammonium hexafluoroaluminate (ammonium fluoroaluminate):

$$(NH_4)_3AlF_6 \rightarrow AlF_3 + 3NH_4F$$

AlF_3 is sparingly soluble in water and is chemically unreactive.

Aluminium chloride can be prepared by passing dry hydrogen chloride or dry chlorine over hot aluminium:

$$2Al + 3Cl_2 \rightarrow 2AlCl_3 \rightarrow Al_2Cl_6$$

$$2Al + 6HCl \rightarrow 3H_2 + 2AlCl_3 \rightarrow Al_2Cl_6$$

Anhydrous aluminium chloride cannot be prepared by dissolving aluminium in hydrochloric acid and then heating the hydrate since the following reaction occurs:

$$2AlCl_3.6H_2O \rightarrow Al_2O_3 + 6HCl + 9H_2O$$

Relative molecular mass determinations in, for example, benzene solution and in the vapour state indicate that the chloride exists as double molecules, Al_2Cl_6 (cf. boron halides which are monomeric). In this way, aluminium atoms complete their octets by dative bonding from two chlorine atoms, giving a roughly tetrahedral arrangement of chlorine atoms about each aluminium atom. These double molecules (1) will persist in the vapour state up to approximately 400 °C, when they start to dissociate to $AlCl_3$ molecules.

Anhydrous aluminium chloride reacts readily with water (it fumes in moist air). A solution of $AlCl_3$ has an acid reaction because of the breakdown of the $[Al(H_2O)_6]^{3+}$ ion (*see* below).

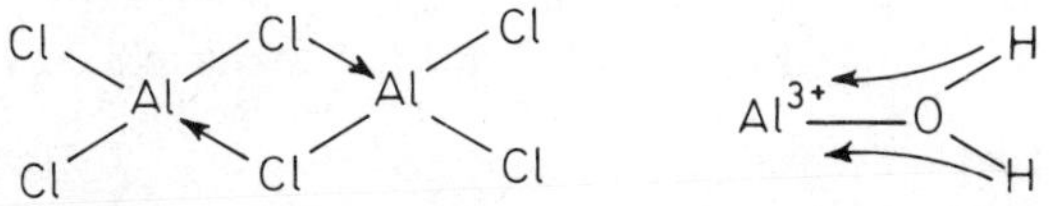

(1) (2)

The structure and properties of aluminium bromide and iodide are very similar to those of the chloride.

12.3.2 ACIDITY OF ALUMINIUM SALTS IN SOLUTION

In all aqueous solutions of aluminium salts, the $[Al(H_2O)_6]^{3+}$ ion is present. The complex cation is acidic because the highly polarizing Al^{3+} ion weakens the O–H bond as indicated in structure (2), (p. 99).

$$[Al(H_2O)_6]^{3+} + H_2O \rightleftharpoons [Al(H_2O)_5(OH)]^{2+} + H_3O^+$$

Consequently, if a weak base (e.g., dilute ammonia solution) is added to this solution, the equilibrium is displaced to the right until the hydroxide is precipitated (i.e., the following sequence occurs):

$$[Al(H_2O)_5(OH)]^{2+} \rightarrow [Al(H_2O)_4(OH)_2]^+ \rightarrow [Al(H_2O)_3(OH)_3]^0 \downarrow$$

Similarly, the hydroxide will be precipitated if the acid is removed from the system by adding, for example, magnesium (to release hydrogen) or a carbonate solution (to release carbon dioxide). However, on adding a strong base (e.g., NaOH) in excess, the precipitated hydroxide dissolves to give the aluminate:

$$[Al(H_2O)_3(OH)_3]^0 \underset{H_3O^+}{\overset{OH^-}{\rightleftharpoons}} [Al(H_2O)_2(OH)_4]^-$$

(Note that the hydroxide not only dissolves in alkalis, but also in acids, i.e., it is amphoteric.)

12.3.3 GENERAL PROPERTIES OF ALUMINIUM

Aluminium is a light metal possessing considerable strength, yet is malleable and ductile. It is not as reactive as its standard electrode potential, –1.66 V (*see* Chapter 9), would imply. This is because of the rapid formation of a thin film of oxide on its surface which prevents further attack by air, water or even dilute acids. The thickness of this film can be deliberately increased by electrolytic oxidation, a process known as ANODIZING. This increases the corrosion resistance of the aluminium.

Aluminium, however, dissolves in moderately concentrated hydrochloric acid, and in hot moderately concentrated sulphuric acid; it is passive to nitric acid (dilute or concentrated) because of the formation of the impervious oxide layer.

$$2Al + 6HCl \rightarrow 2AlCl_3 + 3H_2$$

$$2Al + 6H_2SO_4 \rightarrow Al_2(SO_4)_3 + 6H_2O + 3SO_2$$

Aluminium combines directly with oxygen, sulphur, nitrogen and the halogens if heated to sufficiently high temperatures. Its affinity for oxygen is used in the THERMITE process, where powdered aluminium is mixed with iron(III) oxide and ignited; the iron oxide is reduced to the metal which is then melted by the heat of reaction. Hence this process is used for welding steel (*in situ*).

$$Fe_2O_3 + 2Al \rightarrow Al_2O_3 + 2Fe$$

Aluminium readily dissolves in caustic soda solution to form sodium aluminate.

$$2Al + 2OH^- + 6H_2O \rightarrow 2[Al(OH)_4]^- + 3H_2$$

12.3.4 EXTRACTION OF ALUMINIUM (AND ITS USES)

Aluminium, the most abundant metal in the Earth's crust, is extracted from the ore called bauxite, $Al_2O_3.2H_2O$. This is heated with sodium hydroxide solution under pressure; the aluminium oxide dissolves to form the aluminate and impurities such as iron(III) oxide can be filtered off:

$$Al_2O_3 + 2OH^- + 3H_2O \rightarrow 2[Al(OH)_4]^-$$

From the sodium aluminate solution, aluminium hydroxide is precipitated by 'seeding' with a little freshly prepared aluminium hydroxide:

$$[Al(OH)_4]^- \rightarrow Al(OH)_3 \downarrow + OH^- \text{ (occurs on seeding)}$$

The hydroxide is then heated to give the oxide:

$$2Al(OH)_3 \rightarrow Al_2O_3 + 3H_2O$$

The pure oxide is then dissolved in molten cryolite (Na_3AlF_6) and electrolysed at about 900 °C using graphite electrodes. The following electrode reactions occur:

At cathode (–)	*At anode* (+)
$Al^{3+} + 3e^- \rightarrow Al$	$2O^{2-} - 4e^- \rightarrow O_2$

Aluminium is used for wrapping foodstuffs, for making cooking utensils, overhead cables, and non-tarnishing mirrors; it is used extensively in the car and aircraft industries (both in engines and body panels). Aluminium powder is used in anti-corrosion paints.

12.4 Group IV (C–Pb)

Some of the physical properties of the Group IV elements are shown in *Table 12.5.* Since all the elements have electronic configurations ending in s^2p^2, they are all four electrons short of the noble gas structure. The main group valency is 4 (since an outer *s* electron can be promoted to the empty *p* orbital, giving four unpaired electrons available for bonding (*see* Introduction). Covalency can increase to 6, in complexes, for all elements except carbon [since carbon's (second) outer shell can only hold up to eight electrons]. Changes down the group include:

(1) Elements become more electropositive (because of screening); carbon and silicon are non-metals, germanium has properties of both metals and non-metals (i.e., it is a metalloid), whereas tin and lead are metallic. Oxides of carbon and silicon are acidic, whereas germanium, tin and lead form amphoteric oxides.

(2) The tendency to form bivalent compounds increases.

Table 12.5 shows the large amounts of energy needed to remove four electrons from the Group IV elements. Similarly, the gain of four electrons to form the X^{4-} ion is energetically unfavourable. Consequently, carbon and silicon nearly always form covalent bonds in their compounds. Tin and lead, however,

Table 12.5 Elements of Group IV

Element	*Electronic configuration*	*Ionization energies/kJ mol^{-1}*				*Bond energies*/ kJ mol^{-1}*	*Electronegativities*
		I_1	I_2	I_3	I_4		
C	2.4	1090	2400	4600	6200	346	2.5
Si	2.8.4	790	1600	3200	4400	176	1.8
Ge	2.8.18.4	760	1500	3300	4400	167	1.8
Sn	2.8.18.18.4	710	1400	2900	3900	155	1.8
Pb	2.8.18.32.18.4	720	1500	3100	4100	–	1.8

*Bond energies refer to C–C, Si–Si, Ge–Ge and Sn–Sn.

can both lose two electrons to form the Sn^{2+} and Pb^{2+} ions. Whilst both Ge(II) and Ge(IV) compounds exist, Ge(II), being the least stable, is readily oxidized to Ge(IV) and hence is a strong reducing agent. This occurs to a lesser extent with tin. However, Pb(II) is more stable than Pb(IV) (the reluctance of the outer *s* electrons to enter into bond formation is an example of the INERT PAIR EFFECT); therefore, the latter is a strong oxidizing agent, the lead (IV) being reduced to Pb(II).

12.4.1 GENERAL PROPERTIES

12.4.1.1 Physical

Carbon exists in two allotropic forms, diamond and graphite; their structures and bonding are discussed in Chapter 4, p. 38. Diamond is the hardest natural substance known and is therefore used for cutting and grinding hard substances such as rock, etc. Graphite, unlike diamond, is a conductor of electricity and is therefore particularly useful as an electrode material. It is also used as a lubricant (for reasons, *see* Chapter 4, p. 39).

Neither silicon nor germanium exhibits allotropy. Both have the diamond structure, and both are semiconductors.

Three allotropes of tin are known (shown below, together with their transition temperatures):

$$\underset{\text{(grey)}}{\alpha\text{-tin}} \underset{}{\overset{13.2\,^\circ\text{C}}{\rightleftharpoons}} \underset{\text{(white)}}{\beta\text{-tin}} \overset{161\,^\circ\text{C}}{\rightleftharpoons} \gamma\text{-tin} \overset{232\,^\circ\text{C}}{\rightleftharpoons} \text{liquid tin}$$

α-Tin, or grey tin, has the diamond structure. Above 13.2 °C it is transformed to β-tin, or white tin, which has a metallic nature; γ-tin is also metallic.

Lead does not exhibit allotropy, but exists only in a cubic close packed metallic form.

12.4.1.2 Chemical

12.4.1.2.1 Catenation The chemistry of carbon is dominated by its propensity to form chains (and rings) of carbon atoms (the ability of an element to form bonds between its own atoms to produce long chains is called CATENATION). The tendency to form these chains decreases down the group, as illustrated by the hydrides of the elements (*Table 12.6*). This trend can be explained in terms of the relative bond strengths between the elements (*Table 12.5*); as can be seen, the C–C bonds are particularly strong (since the carbon atoms are small and so the shared electron pair in the bond is close to the nuclei and therefore strongly attracted).

Table 12.6 Catenation in Group IV

Element	*Hydride*
C	Example: poly(ethene), $\left(CH_2 - CH_2 \right)_n$
Si	SiH_4 to Si_6H_{14}
Ge	GeH_4 to Ge_3H_8 (but higher ones have been identified)
Sn	SnH_4, Sn_2H_6
Pb	PbH_4

In the case of silicon, Si–O bonds are much stronger than Si–Si bonds, and many compounds exist which contain the former. Chains containing the –Si–O– repeating unit are present in silicones (used in making water repellants and silicone oils and rubbers), and are prepared as shown in *Figure 12.2.*

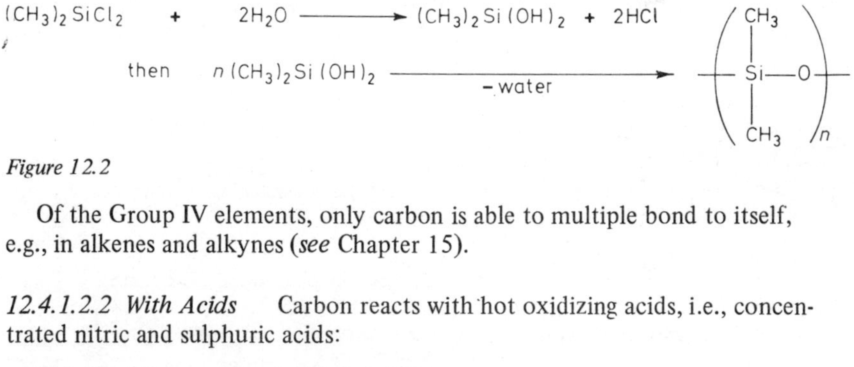

$$(CH_3)_2SiCl_2 + 2H_2O \longrightarrow (CH_3)_2Si(OH)_2 + 2HCl$$

$$\text{then} \quad n(CH_3)_2Si(OH)_2 \xrightarrow[-\text{water}]{} \left(\begin{array}{c} CH_3 \\ | \\ -Si-O- \\ | \\ CH_3 \end{array} \right)_n$$

Figure 12.2

Of the Group IV elements, only carbon is able to multiple bond to itself, e.g., in alkenes and alkynes (*see* Chapter 15).

12.4.1.2.2 With Acids Carbon reacts with hot oxidizing acids, i.e., concentrated nitric and sulphuric acids:

$$C + 2H_2SO_4 \rightarrow CO_2 + 2H_2O + 2SO_2$$

$$C + 4HNO_3 \rightarrow CO_2 + 2H_2O + 4NO_2$$

Silicon is not attacked by acids except hydrofluoric acid:

$$Si + 6HF \rightarrow H_2SiF_6 + 2H_2$$

Germanium, tin and lead are oxidized by concentrated nitric acid:

$$Ge + 4HNO_3 \rightarrow GeO_2 + 2H_2O + 4NO_2$$

$$Sn + 4HNO_3 \rightarrow SnO_2 + 2H_2O + 4NO_2$$

$$3Pb + 8HNO_3 \xrightarrow{\text{Dil. or conc.}} 3Pb(NO_3)_2 + 4H_2O + 2NO$$

Germanium is not attacked by hydrochloric acid, and the dilute acid has little effect on tin and lead. Both tin and lead show little reaction with cold concentrated sulphuric acid, but form tin(II) and lead(II) salts respectively when heated with the acid.

12.4.1.2.3 With Alkalis. Carbon is not affected by alkali. Silicon is easily attacked even by dilute solutions of alkali giving a silicate(IV):

$$Si + 2OH^- + H_2O \rightarrow SiO_3^{2-} + 2H_2$$

Germanium and tin are attacked by hot concentrated alkali solutions forming a germanate(IV) and stannate(IV) respectively, whilst lead forms a plumbate(II) (plumbite):

$$Ge + 2OH^- + H_2O \rightarrow GeO_3^{2-} + 2H_2$$

$$Sn + 2OH^- + H_2O \rightarrow SnO_3^{2-} + 2H_2$$

$$Pb + 2OH^- \rightarrow PbO_2^{2-} + H_2$$

{N.B., an ion such as the stannate(IV) ion is better represented as $[Sn(OH)_6]^{2-}$. Similarly, the aluminate ion is written as $[Al(OH)_4]^-$ rather than AlO_2^-; *see* Group III.}

12.4.2 HYDRIDES

Carbon forms a limitless number of hydrides (details of some of these, and preparations, are given in Chapter 15). SiH_4, GeH_4 and SnH_4 can be prepared by the reaction of the tetrachloride with lithium tetrahydridoaluminate (lithium aluminium hydride) in ethoxyethane (ether), e.g.,

$$SiCl_4 + LiAlH_4 \rightarrow SiH_4 + LiCl + AlCl_3$$

Plumbane, PbH_4, is not very well characterized; it has been formed in minute quantities on treating Mg–Pb alloys with acid.

Table 12.7 Hydrides of Group IV

Hydride	*Formula*	*B.p./°C*	*Properties*
Methane	CH_4	−161	Unaffected by alkali
Silane	SiH_4	−112	Spontaneously flammable in air. Rapidly hydrolysed by OH^-
Germane	GeH_4	−90	Less flammable than silane. Stable in up to 30 per cent alkali
Stannane	SnH_4	−52	Decomposes slowly at room temperature. Stable in up to 15 per cent alkali

Some properties of the hydrides are given in *Table 12.7*, including their tendencies to hydrolyse, e.g.,

$$SiH_4 + 2OH^- + H_2O \rightarrow SiO_3^{2-} + 4H_2$$

Methane's resistance to hydrolysis occurs for several possible reasons. First, the C–H bond is stronger than C–O, whilst the Si–H bond is weaker than Si–O. Secondly carbon has no available *d* orbitals in its (second) outer shell to accept an electron pair from an oxygen atom (Lewis base). Thirdly, because the electronegativities of carbon and hydrogen are 2.5 and 2.1, respectively, the carbon atom in methane has a $\delta-$ charge on it and so has no attraction for Lewis bases (cf. electronegativities of Si, Ge and Sn; *Table 12.5*).

12.4.3 OXIDES

12.4.3.1 Monoxides

All the elements give monoxides. Carbon monoxide is prepared by the dehydration of methanoic (formic) acid using warm concentrated sulphuric acid:

$$HCOOH \xrightarrow{\text{conc. } H_2SO_4} CO + H_2O$$

$$\overset{\times}{\underset{\times}{}}C \vdots O: \quad (\bar{C} \equiv \overset{+}{O}) \qquad\qquad \overset{\times}{\underset{\times}{}}C : \ddot{O}: \quad (C{=}O)$$

(3) (4)

The structure of carbon monoxide is represented as a resonance hybrid of the structures (3) and (4). Important properties of carbon monoxide include

(1) It is a donor molecule and can form compounds called carbonyls with many metals, e.g., tetracarbonylnickel(0) (nickel carbonyl), $Ni(CO)_4$, where donation of two electrons by each carbon atom gives nickel the krypton configuration $[28 + (4 \times 2)]$.
(2) Carbon monoxide is a strong reducing agent, e.g.,

$$PbO + CO \xrightarrow{\text{heat}} Pb + CO_2$$

(3) It adds to chlorine in sunlight (or in the dark in the presence of charcoal) to give carbonyl chloride (phosgene), $COCl_2$.
(4) It reacts with sodium hydroxide solution above 150 °C (under pressure) to give sodium methanoate (formate):

$$CO + NaOH \rightarrow HCOONa$$

Silicon(II) oxide is formed by the reduction of silicon(IV) oxide (silica) by silicon at high temperature. Germanium(II) oxide [prepared by reduction of Ge(IV) in solution] is unstable and tends to DISPROPORTIONATE* on heating:

$$2Ge(II) \rightarrow Ge(IV) + Ge$$

Tin(II) oxide is prepared by heating the ethanedioate (oxalate):

$$SnC_2O_4 \rightarrow SnO + CO + CO_2$$

whereas lead(II) oxide can be prepared by heating the nitrate:

$$2Pb(NO_3)_2 \longrightarrow \underset{\text{(yellow or red)}}{2PbO} + 4NO_2 + O_2$$

Both SnO and PbO are amphoteric, giving tin(II) and lead(II) salts, respectively, in acid and the stannate(II) (stannite) and plumbate(II) (plumbite), respectively, in alkali.

12.4.3.2 Dioxides

All the elements give dioxides; CO_2 and SiO_2 are acidic whilst GeO_2, SnO_2 and PbO_2 are amphoteric.

*Disproportionation is the simultaneous oxidation and reduction of a species.

Carbon dioxide is prepared by treating a carbonate with acid or by heating hydrogencarbonates or most carbonates. Its shape is linear (O=C=O). It is an acidic gas, forming weak carbonic acid in water:

$$H_2O + CO_2 \rightleftharpoons H_2CO_3$$

Silicon(IV) oxide (silica or silicon dioxide) occurs naturally as quartz or kieselguhr. It consists of silicon atoms linked tetrahedrally to four oxygen atoms to make a 'giant' molecule (5). When prepared by hydrolysing SiF_4 or $SiCl_4$, it is

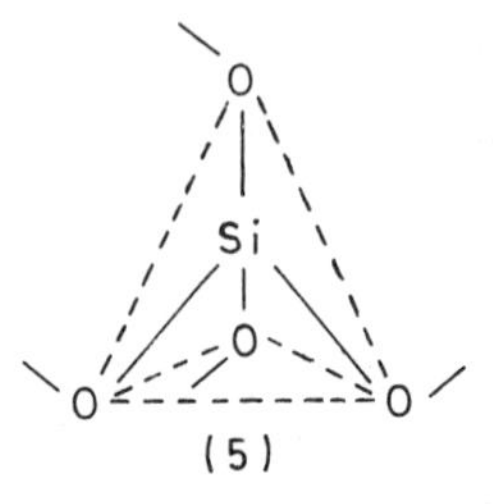

(5)

produced as a gel which can be made anhydrous on heating. This solid will absorb water if exposed to a moist atmosphere, and consequently silica gel is an important drying agent. Silica is attacked only by hydrofluoric acid, and by alkali to give silicates:

$$SiO_2 + 2OH^- \rightarrow SiO_3^{2-} + H_2O$$

Germanium(IV) oxide is amphoteric, dissolving in concentrated hydrochloric acid to give the Ge(IV) salt, and in alkali to give the germanate(IV).

Tin(IV) oxide, like germanium(IV) oxide, can be obtained by heating the element in oxygen. It is amphoteric, reacting with acids to give tin(IV) salts and with alkalis to give the stannate(IV) [strictly, the hexahydroxostannate(IV)]:

$$SnO_2 + 2OH^- + 2H_2O \rightarrow [Sn(OH)_6]^{2-}$$

Lead(IV) oxide is prepared by adding dilute nitric acid to dilead(II) lead(IV) oxide (red lead), Pb_3O_4:

$$Pb_3O_4 + 4HNO_3 \rightarrow 2Pb(NO_3)_2 + \underset{\text{(brown)}}{PbO_2\downarrow} + 2H_2O$$

Lead(IV) oxide evolves oxygen on heating:

$$2PbO_2 \rightarrow 2PbO + O_2$$

and it is amphoteric; it oxidizes concentrated hydrochloric acid, and gives a plumbate(IV) [strictly, the hexahydroxoplumbate(IV)] in alkali:

$$PbO_2 + 4HCl \rightarrow PbCl_2 + 2H_2O + Cl_2$$

$$PbO_2 + 2OH^- + 2H_2O \rightarrow [Pb(OH)_6]^{2-}$$

12.4.3.3 Dilead(II) Lead(IV) Oxide

This oxide, Pb_3O_4, is prepared by heating lead(II) oxide in air at about 400 °C:

$$6PbO + O_2 \rightarrow 2Pb_3O_4$$

It evolves oxygen on heating:

$$2Pb_3O_4 \rightarrow 6PbO + O_2$$

12.4.4 HALIDES

12.4.4.1 Dihalides

Germanium, tin and lead form dihalides (with some ionic character) although methods of preparation differ.

Germanium dihalides can be made by heating germanium(IV) halides with germanium:

$$GeCl_4 + Ge \rightarrow 2GeCl_2$$

Tin(II) fluoride is prepared by dissolving tin(II) oxide in hydrofluoric acid, whilst tin(II) chloride is prepared by dissolving tin in concentrated hydrochloric acid when crystals of $SnCl_2.2H_2O$ are deposited on cooling:

$$Sn + 2HCl \rightarrow SnCl_2 + H_2$$

Tin(II) chloride is a reducing agent, e.g.,

$$2Fe^{3+} + Sn^{2+} \rightarrow 2Fe^{2+} + Sn^{4+}$$

$$2HgCl_2 + SnCl_2 \rightarrow Hg_2Cl_2\downarrow + SnCl_4$$

then in excess of $SnCl_2$:

$$Hg_2Cl_2 + SnCl_2 \rightarrow 2Hg\downarrow + SnCl_4$$

Lead dihalides are obtained by double decomposition, e.g.,

$$Pb^{2+} + 2Cl^- \rightarrow PbCl_2\downarrow \text{ (white)}$$
$$Pb^{2+} + 2I^- \rightarrow PbI_2\downarrow \text{ (yellow)}$$

The chloride dissolves in concentrated HCl to form the $[PbCl_4]^{2-}$ complex ion.

12.4.4.2 Tetrahalides

All the Group IV tetrahalides are known except for $PbBr_4$ and PbI_4 [Br_2 and I_2 are unable to oxidize the Pb(II) to Pb(IV)]. They tend to be covalent; tin(IV) and lead(IV) fluorides are solids having appreciable ionic character, however.

Tetrachloromethane (carbon tetrachloride) is made by chlorinating carbon disulphide in the presence of an iron catalyst:

$$CS_2 + 3Cl_2 \rightarrow CCl_4 + S_2Cl_2$$

The tetrachlorides of the remaining elements are very susceptible to hydrolysis; they are therefore made under anhydrous conditions by direct combination of elements, except for lead which forms the dichloride. Lead(IV) chloride can be prepared from cold concentrated hydrochloric acid and lead(IV) oxide:

$$PbO_2 + 4HCl \xrightarrow{0\,°C} PbCl_4 + 2H_2O$$

(If the temperature is allowed to rise, PbO_2 oxidizes the hydrochloric acid to chlorine; *see* before.) The lead(IV) chloride readily evolves chlorine by the reversible reaction

$$PbCl_4 \rightleftharpoons PbCl_2 + Cl_2$$

again showing the greater stability of Pb(II). Hence, if chlorine is passed into a cold suspension of lead(II) chloride in hydrochloric acid, lead(IV) chloride is formed; if ammonium chloride is added, the complex salt ammonium hexachloroplumbate(IV) is precipitated:

$$PbCl_4 + 2NH_4Cl \rightarrow (NH_4)_2PbCl_6 \downarrow \text{ (yellow)}$$

On filtering and adding concentrated sulphuric acid, $PbCl_4$ separates as an oily yellow liquid:

$$(NH_4)_2PbCl_6 + H_2SO_4 \rightarrow (NH_4)_2SO_4 + PbCl_4 + 2HCl$$

Other complexes can be formed from the tetrahalides, including the hexachlorostannate(IV) ion, $[SnCl_6]^{2-}$, made by adding tin(IV) chloride to concentrated hydrochloric acid.

The tetrachlorides of Group IV are all covalent liquids and, except CCl_4, fume in air because of hydrolysis, e.g.,

$$SnCl_4 + 4H_2O \rightarrow SnO_2.2H_2O + 4HCl$$

The resistance of CCl_4 (and the other carbon tetrahalides) to hydrolysis is because of carbon's maximum covalency of 4 (no *d* orbitals; *see* also the hydrides).

12.5 Group V (Nitrogen and Phosphorus)

Nitrogen and phosphorus (atomic numbers 7 and 15 respectively) are the first two members of Group V. Their electronic configurations are $1s^2 2s^2 2p^3$ and $[Ne]3s^2 3p^3$ respectively, i.e., both have three unpaired electrons. They are both typical non-metals, having acidic oxides. Nitrogen can (1) share three electrons to give a covalency of 3, e.g., NH_3; (2) acquire three electrons to form the nitride ion, N^{3-}, e.g., Li_3N (nitrogen is one of the most electronegative elements); (3) share three electrons and donate the lone pair to an acceptor atom or molecule (or ion), to give a maximum covalency of 4, e.g., NH_4^+ or $H_3N \rightarrow AlCl_3$. A range of oxidation states occurs with nitrogen:

−3	−2	−1	0	+1	+2	+3	+4	+5
NH_3	N_2H_4	NH_2OH	N_2	N_2O	NO	NO_2^-	N_2O_4	NO_3^-

Phosphorus, however, can increase its covalency to 5, e.g., PCl_5, (by promotion of a 3*s* electron to a 3*d* orbital, giving five unpaired electrons available for sharing), or even 6, e.g., PF_6^- or PCl_6^- (again, because of the availability of the 3*d* orbitals).

Nitrogen is a colourless, odourless, tasteless gas, only very slightly soluble in water, and occurs to the extent of about 78 per cent by volume in air. It exists as diatomic molecules containing a triple bond, and is chemically inert because of the molecule's very high dissociation energy (945 kJ mol^{-1}; cf. halogens, *see* Group VII). Nitrogen can be prepared in the laboratory by heating a concentrated solution of ammonium nitrite*,

$$NH_4NO_2 \rightarrow N_2 + 2H_2O$$

or by heating ammonium dichromate(VI):

$$(NH_4)_2Cr_2O_7 \rightarrow Cr_2O_3 + 4H_2O + N_2$$

Nitrogen is obtained industrially by the fractionation of liquid air. It is used in the laboratory as an inert atmosphere; liquid nitrogen is used as a cooling agent. Nitrogen is used industrially in the Haber process (*see* below and also p. 63).

Table 12.8 Allotropes of phosphorus

White	*Red*
White waxy solid	Red brittle powder
Density 1.82 g cm^{-3}	Density 2.20 g cm^{-3}
Exists as tetrahedral P_4 units	Macromolecular
Melting point: 44 °C	Melting point: 597 °C (under pressure)
Soluble in carbon disulphide, benzene, etc.	Insoluble in these solvents
Ignites in damp air at about 30 °C	Ignites at about 300 °C
Ignites in chlorine	Ignites in chlorine when heated
Reacts with hot alkali to form phosphine	No reaction with alkali

Phosphorus occurs in nature as salts of phosphoric(V) acid, the most common being apatite, $3Ca_3(PO_4)_2.CaF_2$. Phosphorus exhibits allotropy, the white and red allotropes being the two common ones. Some of their properties are given in *Table 12.8.* Phosphorus was used in rat poisons, and is used in the match and fertilizer industries.

12.5.1 OXIDES

12.5.1.1 Dinitrogen Oxide (Nitrous Oxide).

This is prepared by the careful heating of ammonium nitrate:

$$NH_4NO_3 \rightarrow N_2O + 2H_2O$$

It is a colourless gas, slightly soluble in water, giving a neutral solution. It is chemically unreactive; it decomposes into its elements on heating. Hence, a substance burning with a flame hot enough to decompose it will continue to burn in it because of the liberated oxygen. Its structure is as shown in (6).

$$\overset{-}{N}=\overset{+}{N}=O \longleftrightarrow N\equiv\overset{+}{N}-\overset{-}{O}$$

(6)

The molecule is linear, and in this sense it resembles carbon dioxide, with which it is ISOELECTRONIC (has the same number of electrons).

*Ammonium nitrite can be prepared by adding ammonium chloride to sodium nitrite.

12.5.1.2 *Nitrogen Oxide (Nitric Oxide)*

This is prepared by the action of a mixture of equal volumes of concentrated nitric acid and water on copper turnings:

$$3Cu + 8HNO_3 \rightarrow 3Cu(NO_3)_2 + 4H_2O + 2NO$$

It is a colourless gas, only slightly soluble in water. It is not so easily dissociated into its elements as dinitrogen oxide, and so will only support the combustion of vigorously burning substances such as phosphorus (but not charcoal or sulphur). It is more reactive than dinitrogen oxide because it possesses a single unpaired electron (and is therefore said to be PARAMAGNETIC, i.e., attracted by a magnetic field), as shown by structure (7).

$$\overset{\times\times}{\underset{\times}{\times}}N\overset{\times}{\underset{\times}{:}}\overset{\bullet\bullet}{\underset{\bullet\bullet}{O}}\quad (\overset{\times}{\underset{\times}{\times}}N{=}\ddot{O}:)$$

(7)

Nitrogen oxide can lose an electron to give the nitrosyl cation (nitrosonium ion), NO^+, e.g., the salt $(NO)^+(BF_4)^-$, nitrosyl tetrafluoroborate, or can gain an electron, e.g.,

$$Na + NO \xrightarrow{\text{liquid ammonia}} Na^+NO^- \text{ (sodium nitrosyl)}$$

Nitrogen oxide can also donate electrons. For example, in the brown ring test for nitrates [if concentrated sulphuric acid is run down the side of a test tube into a mixture of iron(II) sulphate solution and a nitrate solution, a brown ring forms at the solution/acid junction], the brown compound is formed by a water molecule in the $[Fe(H_2O)_6]^{2+}$ complex being replaced by NO to give $[Fe(H_2O)_5(NO)]^{2+}$.

Nitrogen oxide combines with oxygen spontaneously to give nitrogen dioxide, and with chlorine to give nitrosyl chloride:

$$2NO + O_2 \rightarrow 2NO_2$$

$$2NO + Cl_2 \rightarrow 2NOCl$$

12.5.1.3 *Nitrogen Dioxide and Dinitrogen Tetraoxide*

Nitrogen dioxide is prepared by heating lead nitrate; the dioxide can be condensed using a freezing mixture:

$$2Pb(NO_3)_2 \rightarrow 2PbO + 4NO_2 + O_2$$

At temperatures below $-10\,^\circ C$, it exists as colourless crystals of the dimer, N_2O_4. On raising the temperature, it forms a pale brown liquid which, at $22.4\,^\circ C$, boils. On heating further, the gas becomes red-brown and then almost black at $150\,^\circ C$, when dissociation to NO_2 is complete. Further increase in temperature results in a loss of colour as the nitrogen dioxide decomposes into nitrogen oxide and oxygen:

$$N_2O_4 \underset{\text{cool}}{\overset{\text{heat}}{\rightleftharpoons}} 2NO_2 \underset{\text{cool}}{\overset{\text{heat}}{\rightleftharpoons}} 2NO + O_2$$

Nitrogen dioxide is paramagnetic, having an unpaired electron. Since both N–O bonds are of equal length, its structure must be a resonance hybrid of the structures (8)–(11). When it dimerizes, the unpaired electrons become paired

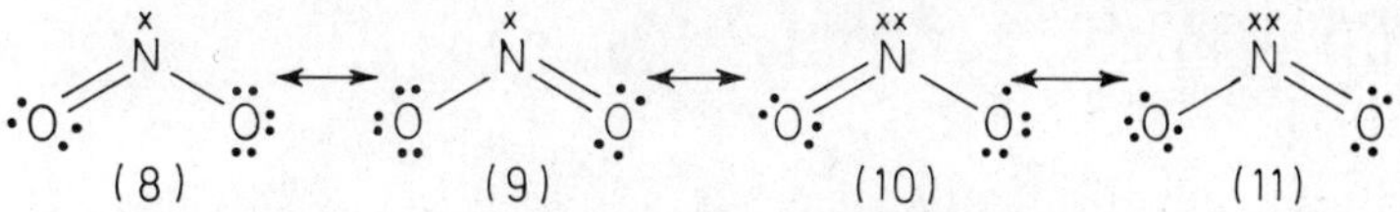

(8) (9) (10) (11)

(and the molecule is said to be DIAMAGNETIC, i.e., repelled by a magnetic field). The structure (12) is planar:

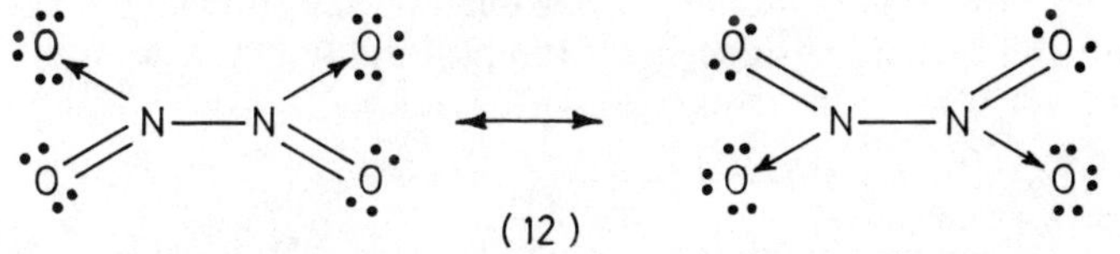

(12)

Nitrogen dioxide dissolves in water to give a mixture of nitrous and nitric acids:

$$2NO_2 + H_2O \rightarrow HNO_2 + HNO_3$$

but since the nitrous acid decomposes rapidly at room temperature by the equation:

$$3HNO_2 \rightarrow HNO_3 + 2NO + H_2O$$

the overall reaction is:

$$H_2O + 3NO_2 \rightarrow 2HNO_3 + NO \text{ (oxidized by air to } NO_2\text{)}$$

With alkalis, nitrogen dioxide gives the nitrite and nitrate:

$$2OH^- + 2NO_2 \rightarrow NO_3^- + NO_2^- + H_2O$$

Nitrogen dioxide is a powerful oxidizing agent (hence phosphorus, charcoal and sulphur burn in it to yield their oxides and nitrogen). It oxidizes hydrogen sulphide to sulphur:

$$H_2S + NO_2 \rightarrow NO + H_2O + S$$

However, it is also a reducing agent, decolourizing acidified manganate(VII), and itself being oxidized to the nitrate ion:

$$2MnO_4^- + 10NO_2 + 2H_2O \rightarrow 2Mn^{2+} + 4H^+ + 10NO_3^-$$

12.5.1.4 Phosphorus(III) Oxide (Phosphorus Trioxide)

Phosphorus(III) oxide was once thought to be P_2O_3, but is now known (from relative molecular mass measurements in solution and in the vapour state) to be the dimer, P_4O_6. It is made by burning white phosphorus in a limited amount of air, forming a white solid:

$$P_4 + 3O_2 \rightarrow P_4O_6$$

It reacts rapidly when heated with oxygen to give phosphorus(V) oxide (phosphorus pentoxide), and with water to give phosphonic acid (phosphorous acid):

$$P_4O_6 + 2O_2 \rightarrow P_4O_{10}$$

$$P_4O_6 + 6H_2O \rightarrow 4H_3PO_3$$

12.5.1.5 Phosphorus(V) Oxide (Phosphorus Pentoxide)

Phosphorus(V) oxide, once given the formula P_2O_5 but now known to be the dimer P_4O_{10}, is prepared by burning phosphorus in a plentiful supply of air to give a white solid:

$$P_4 + 5O_2 \rightarrow P_4O_{10}$$

With water, followed by boiling, phosphoric(V) acid (orthophosphoric acid) is formed:

$$P_4O_{10} + 2H_2O \rightarrow 4HPO_3$$

then

$$HPO_3 + H_2O \rightarrow H_3PO_4$$

12.5.2 HYDRIDES

12.5.2.1 Ammonia

Ammonia is prepared in the laboratory by heating an ammonium salt with an alkali, e.g.,

$$2NH_4Cl + Ca(OH)_2 \rightarrow CaCl_2 + 2NH_3 + 2H_2O$$

It is also obtained when an ionic nitride is hydrolysed by water:

$$N^{3-} + 3H_2O \rightarrow 3OH^- + NH_3$$

Industrially, it is made by the Haber process.

It is a colourless gas with a characteristic pungent smell. Its shape is based on the tetrahedron [*see Figure 2.2(a)*]. It is extremely soluble in water, to which it can hydrogen bond. It can also abstract a proton from a water molecule by using its lone pair (i.e., ammonia is acting as a Lewis base), so two competing factors are operating:

$$:NH_3 + H_2O \rightleftharpoons H_3N:\text{------}H{-}O{-}H \text{ (written as } NH_3.H_2O)$$

$$NH_3 + H_2O \rightleftharpoons NH_4^+ + OH^-$$

The overall effect is now:

$$NH_3.H_2O \rightleftharpoons NH_4^+ + OH^-$$

and so the ammonia solution ('loosely' called ammonium hydroxide) is said to act as a weak base owing to its slight dissociation. Hence, ammonia solution can precipitate metal hydroxides from solutions of their salts (provided their solubility products are exceeded – *see* Chapter 8). However, it sometimes happens that these hydroxides redissolve in excess ammonia solution because of the formation of soluble ammine complexes, e.g., $[Cu(NH_3)_4(H_2O)_2]^{2+}$ (*see* Chapter 13).

Liquid ammonia is slightly ionized:

$$2NH_3 \rightleftharpoons NH_4^+ + NH_2^- \text{ (cf. } 2H_2O \rightleftharpoons H_3O^+ + OH^-)$$

Hence, ammonium salts and amides dissolved in liquid ammonia behave as acids and bases respectively, e.g.,

$$\underset{\text{acid}}{NH_4Cl} + \underset{\text{base}}{KNH_2} \rightarrow \underset{\text{salt}}{KCl} + \underset{\text{solvent}}{2NH_3}$$

12.5.2.2 Phosphine

Phosphine, PH_3, may be prepared by the action of a strong alkali on white phosphorus or on phosphonium iodide:

$$P_4 + 3KOH + 3H_2O \rightarrow 3KH_2PO_2 + PH_3$$

where KH_2PO_2 is potassium phosphinate (hypophosphite).

$$PH_4I + KOH \rightarrow KI + H_2O + PH_3$$

Phosphine is a colourless, poisonous gas with an unpleasant smell. It is slightly soluble in water, but the equilibrium

$$PH_3 + H_2O \rightleftharpoons PH_4^+ + OH^-$$

is well to the left, i.e., the solution is only weakly basic. Its lower solubility, compared with that of ammonia, is because phosphorus is not sufficiently electronegative to enable hydrogen bonding to occur with water molecules. Similarly, the presence of hydrogen bonding in ammonia, but not in phosphine, explains why the latter has the lower boiling point even though it is heavier than ammonia (ammonia, b.p. –33 °C; phosphine, b.p. –90 °C).

12.5.3 CHLORIDES

Nitrogen forms a trichloride when excess chlorine reacts with ammonia. It is a covalent and highly explosive yellow oil which is readily hydrolysed to ammonia and chloric(I) acid (hypochlorous acid):

$$2NH_3 + 6Cl_2 \rightarrow 2NCl_3 + 6HCl$$

$$NCl_3 + 3H_2O \rightarrow NH_3 + 3HOCl$$

Phosphorus forms a trichloride and a pentachloride by direct combination; which one is formed depends on the relative amounts of phosphorus and chlorine present:

$$P_4 + 6Cl_2 \rightarrow 4PCl_3$$

$$PCl_3 + Cl_2 \rightarrow PCl_5$$

The trichloride is a covalent liquid, the pentachloride is a solid having an ionic lattice built up of $[PCl_4]^+$ and $[PCl_6]^-$ ions. Both are hydrolysed by water:

$$PCl_3 + 3H_2O \rightarrow H_3PO_3 + 3HCl$$

$$PCl_5 + H_2O \rightarrow POCl_3 + 2HCl$$

then

$$POCl_3 + 3H_2O \rightarrow H_3PO_4 + 3HCl$$

$POCl_3$ is phosphorus trichloride oxide (phosphorus oxychloride).

12.5.4 NITRIC ACID AND ITS REACTIONS

Nitric acid is prepared in the laboratory by heating equal masses of potassium nitrate and concentrated sulphuric acid. Its structure is shown on p. 23.

$$H_2SO_4 + KNO_3 \rightarrow KHSO_4 + HNO_3$$

It is a powerful oxidizing agent. The reduction products of these redox reactions depend on the pH (or concentration of acid used) and the nature of the material being oxidized. For example, with a mixture of equal volumes of concentrated nitric acid and water on copper, nitrogen oxide (nitric oxide) is produced, but if concentrated nitric acid is used, nitrogen dioxide is formed:

$$3Cu + 8HNO_3 \rightarrow 3Cu(NO_3)_2 + 2NO + 4H_2O$$

$$Cu + 4HNO_3 \rightarrow Cu(NO_3)_2 + 2NO_2 + 2H_2O$$

Metals which do liberate hydrogen from dilute acids, e.g., zinc, magnesium, can react with nitric acid to give dinitrogen oxide (nitrous oxide):

$$4Zn + 10HNO_3 \rightarrow 4Zn(NO_3)_2 + N_2O + 5H_2O$$

However, with very dilute nitric acid on magnesium, hydrogen can be formed:

$$Mg + 2HNO_3 \rightarrow Mg(NO_3)_2 + H_2$$

Nitric acid can oxidize hydrogen sulphide to sulphur, iodides to iodine and iron(II) salts to iron(III).

$$3H_2S + 2HNO_3 \rightarrow 4H_2O + 2NO + 3S$$

$$2HI + HNO_3 \rightarrow HNO_2 + I_2 + H_2O$$

$$3Fe^{2+} + NO_3^- + 4H^+ \rightarrow 3Fe^{3+} + NO + 2H_2O$$

12.5.5 NITROGEN IN INDUSTRY

Nitrogen (from the atmosphere) is used in large quantities for the manufacture of ammonia by the Haber process (*see* p. 63):

$$N_2 + 3H_2 \rightleftharpoons 2NH_3$$

The necessary hydrogen is taken from cracking processes, in which it is a by-product, and so many ammonia plants are now situated at or near oil refineries. The yield of ammonia is about 12–15 per cent; this is condensed out and the unused gases are recycled.

A large amount of ammonia is used for the manufacture of nitric acid. Here, the ammonia is oxidized by air using a platinum–rhodium catalyst at about 850 °C:

$$4NH_3 + 5O_2 \rightarrow 4NO + 6H_2O$$

The mixture is cooled, when the nitrogen oxide (nitric oxide) reacts with air to form nitrogen dioxide, followed by water and more air to produce nitric acid:

$$2NO + O_2 \rightarrow 2NO_2$$

$$4NO_2 + 2H_2O + O_2 \rightarrow 4HNO_3$$

Nitric acid is used for nitrating aromatic hydrocarbons (p. 160) to form nitro-compounds, which can in turn be reduced to amines for use in the dye industry (p. 186). Amines are also used for making polyamides (p. 193).

Nitrogen is an essential constituent of all living organisms, both plant and animal, and is present in the form of proteins. Soluble nitrogen compounds, therefore, are used in vast quantities for the fertilizer industry, e.g., ammonium sulphate, nitrate and phosphate(V). However, not all fertilizers are nitrogenous compounds, e.g., superphosphate[$Ca(H_2PO_4)_2 + CaSO_4$], made by adding concentrated sulphuric acid to calcium phosphate(V).

12.6 Group VI (Oxygen and Sulphur)

Oxygen and sulphur are the first two elements in Group VI, having electronic configurations $1s^2 2s^2 2p^4$ and $[Ne]3s^2 3p^4$ respectively. Having six electrons in their outer shells, both oxygen and sulphur can acquire two more electrons to form the oxide and sulphide ions (O^{2-} and S^{2-}) respectively, so attaining noble gas structure. Most metal oxides are wholly or partly ionic (soluble ionic oxides, e.g., Na_2O, in water undergo hydrolysis to give the hydroxide ion). Sulphides of the more electropositive metals are ionic; other metal sulphides have some covalent character. Both oxygen and sulphur combine with non-metals giving a covalency of 2, e.g., H_2O and H_2S. Water and its derivatives, e.g., ethoxyethane (ether), $(C_2H_5)_2O$, can act as Lewis bases because of their lone pairs, e.g., $H_2O + H^+ \rightarrow H_3O^+$, giving oxygen a covalency of 3. Oxygen can, in fact, have a covalency maximum of 4, but this is rare. In contrast, sulphur can increase its covalency to 6 because of electron promotion to *d* orbitals (*see* Introduction, p. 10). Both oxygen and sulphur can form multiple bonds, e.g., $>C{=}O$, $O{=}S{=}O$, etc. Oxygen has little tendency to form O–O bonds, and when it does, e.g., in peroxides, they are readily broken by heat. Sulphur, however, demonstrates its power of catenation by forming the S_8 molecule.

Oxygen comprises about 20 per cent of the atmosphere. It can be prepared in the laboratory by the thermal decomposition of various oxides and oxo-salts, e.g., HgO, $KClO_3$ (catalysed by MnO_2), and $KMnO_4$. The catalytic decomposition of hydrogen peroxide (using MnO_2) is also a common method:

$$2KClO_3 \rightarrow 2KCl + 3O_2$$

$$2KMnO_4 \rightarrow K_2MnO_4 + MnO_2 + O_2$$

$$2H_2O_2 \rightarrow 2H_2O + O_2$$

Oxygen exhibits allotropy, the two forms being dioxygen, O_2, and trioxygen (ozone), O_3. Trioxygen can be formed when air is subjected to an electric discharge (it is formed in the upper atmosphere). It is diamagnetic and has the structure (13). Oxygen is a supporter of combustion, many metals and non-metals burning in it to give oxides. Its ability to stabilize high oxidation states, e.g., PbO_2, SO_3, Cl_2O_7, etc., is because of its small size and high electronegativity (which is 3.5).

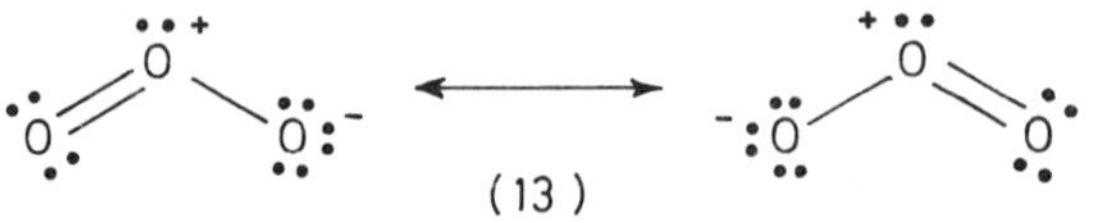

(13)

Sulphur occurs as deposits of free sulphur (found in America, Japan and Sicily); some is also obtained from the purification of petroleum and natural gas. It also occurs in sulphides, e.g., galena, PbS; zinc blende, ZnS; iron pyrites, FeS_2, and in sulphates, e.g. gypsum or anhydrite, $CaSO_4$. Sulphur exhibits allotropy. Rhombic sulphur is stable at room temperature and monoclinic

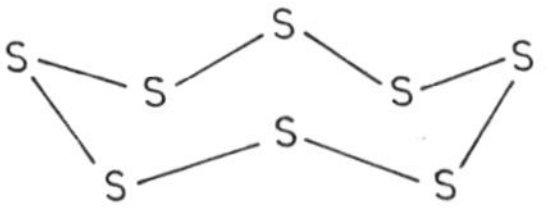

(14)

sulphur is stable above 96 °C; both these allotropes contain S_8 molecules with rings of eight sulphur atoms (14). Other forms of sulphur include plastic and amorphous sulphur.

Uses of sulphur include: manufacture of matches and fireworks; insecticides; vulcanizing rubber; the manufacture of sulphuric acid (*see* later).

12.6.1 HYDRIDES

12.6.1.1 Water

Some of the properties of water include: maximum density at 4 °C (*see* p. 25), no smell or taste (cf. other hydrides of the group); high enthalpies (latent heats) of fusion and vaporization; liquid at normal temperatures (cf. other hydrides of the group); high dielectric constant.

The high enthalpies of fusion and vaporization, and the high boiling point of water, are a direct consequence of hydrogen bonding (*see* p. 25). Water is a highly effective solvent; this is because

(1) it is polarized, and has lone pairs, and can therefore solvate ions (*see* p. 26), and
(2) it can form hydrogen bonds to many (polar) covalent substances, e.g., alcohols.

Although pure water is virtually a non-conductor of electricity, it is slightly dissociated,

$$2H_2O \rightleftharpoons H_3O^+ + OH^-$$

and at 25 °C the oxonium ion (H_3O^+) and hydroxide ion concentrations are both 10^{-7} mol l^{-1} (*see* also Chapter 8, p. 67).

Many substances undergo hydrolysis reactions with water, where hydrolysis can mean

(1) the direct reaction of water with a substance, e.g., the hydrolysis of acid chlorides (*see* p. 178) because of lone pair attack by water, or
(2) the dissociation of water in a hydrate shell to give oxonium ions, e.g.,

$$[Al(H_2O)_6]^{3+} + H_2O \rightleftharpoons [Al(H_2O)_5(OH)]^{2+} + H_3O^+$$

12.6.1.2 Hydrogen Sulphide

This is a colourless gas, having a most unpleasant smell. Its low boiling point compared with that of water (*see Table 3.1*) is because of the absence of hydrogen bonding (the electronegativity of sulphur is 2.5; cf. oxygen 3.5). Lack of hydrogen bonding with water molecules themselves also explains its low solubility in water.

Hydrogen sulphide can be prepared by adding dilute hydrochloric acid to iron(II) sulphide, or by adding cold water to aluminium sulphide (which yields a purer product):

$$FeS + 2HCl \rightarrow FeCl_2 + H_2S$$

$$Al_2S_3 + 6H_2O \rightarrow 2Al(OH)_3 + 3H_2S$$

Hydrogen sulphide burns in air with a blue flame:

$$2H_2S + 3O_2 \rightarrow 2SO_2 + 2H_2O$$

It is also a weak acid in solution, dissolving in alkalis to give sulphides and hydrogensulphides:

$$2OH^- + H_2S \rightleftharpoons S^{2-} + 2H_2O$$

$$OH^- + H_2S \rightleftharpoons HS^- + H_2O$$

Hydrogen sulphide is a powerful reducing agent, as illustrated by the following examples:

(*a*) (moist) $Cl_2 + H_2S \rightarrow 2HCl + S$

(*b*) (moist) $SO_2 + 2H_2S \rightarrow 2H_2O + 3S$

(*c*) $H_2SO_4 + H_2S \rightarrow SO_2 + 2H_2O + S$

(*d*) $Cr_2O_7^{2-} + 8H^+ + 3H_2S \rightarrow 2Cr^{3+} + 7H_2O + 3S$

$Cr_2O_7^{2-}$ is the orange dichromate(VI) ion; chromium(III) ion, Cr^{3+}, is green.

(*e*) $2MnO_4^- + 5H_2S + 6H^+ \rightarrow 2Mn^{2+} + 8H_2O + 5S$

MnO_4^- is the purple manganate(VII) (permanganate) ion; Mn^{2+}, the manganese(II) ion is, strictly speaking, pink, but appears colourless when in low concentration.

(*f*) $2Fe^{3+} + H_2S \rightarrow 2Fe^{2+} + 2H^+ + S$

Because most metallic sulphides are insoluble in water, many are therefore precipitated when hydrogen sulphide is passed through solutions containing these metal ions, e.g.,

$$Pb^{2+} + H_2S \rightarrow PbS\downarrow \text{ (black)} + 2H^+$$

$$Cu^{2+} + H_2S \rightarrow CuS\downarrow \text{ (black)} + 2H^+$$

The test for hydrogen sulphide is the blackening of filter paper which has been dipped in lead ethanoate (acetate) solution.

12.6.2 OXIDES OF SULPHUR (INCLUDING OXOACIDS AND OXOANIONS)

Sulphur dioxide is a pungent smelling gas, easily liquefied under pressure, and can be prepared by the action of hot concentrated sulphuric acid on copper, or by the action of an acid on a sulphite or hydrogensulphite:

$$Cu + 2H_2SO_4 \rightarrow CuSO_4 + 2H_2O + SO_2$$

$$Na_2SO_3 + 2HCl \rightarrow 2NaCl + H_2O + SO_2$$

Its structure is (15). Sulphur dioxide is an acidic oxide, dissolving readily in

(15)

water to give sulphurous acid, H_2SO_3. It dissolves in alkalis to give hydrogensulphites and sulphites:

$$OH^- + SO_2 \rightarrow HSO_3^-$$

$$HSO_3^- + OH^- \rightarrow SO_3^{2-} + H_2O$$

Sulphurous acid, and solutions of hydrogensulphites and sulphites, are reducing agents, e.g.,

$$2MnO_4^- + 5SO_3^{2-} + 6H^+ \rightarrow 2Mn^{2+} + 5SO_4^{2-} + 3H_2O$$

$$I_2 + SO_3^{2-} + H_2O \rightarrow 2I^- + SO_4^{2-} + 2H^+$$

$$MnO_2 + 2SO_3^{2-} + 4H^+ \rightarrow \underset{\text{(dithionate)}}{S_2O_6^{2-}} + Mn^{2+} + 2H_2O$$

On boiling a solution of a sulphite with sulphur, a thiosulphate(VI) (thiosulphate) is formed:

$$SO_3^{2-} + S \rightarrow S_2O_3^{2-}$$

The thiosulphate(VI) is oxidized by iodine to the tetrathionate (and this reaction is extremely important in volumetric analysis):

$$2S_2O_3^{2-} + I_2 \rightarrow S_4O_6^{2-} + 2I^-$$

and also reacts with acid to give sulphur dioxide and sulphur:

$$S_2O_3^{2-} + 2H^+ \rightarrow SO_2 + S + H_2O$$

Sulphur(VI) oxide (sulphur trioxide) can be prepared in the laboratory by passing a mixture of sulphur dioxide and oxygen over a heated catalyst of platinized asbestos. It can be condensed as a white solid in a cooled receiver. Its structure is (16), a planar molecule with all bond angles 120 degrees. It is a

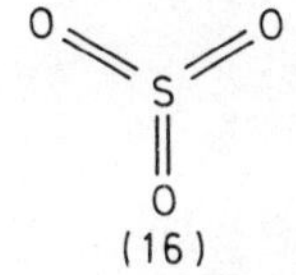

(16)

powerful, acidic oxide, which fumes in moist air and reacts explosively with water to form sulphuric acid:

$$SO_3 + H_2O \rightarrow H_2SO_4$$

12.6.3 SULPHURIC ACID (ITS PROPERTIES AND USES)

The industrial manufacture of sulphuric acid (the Contact process) involves the catalytic oxidation of sulphur dioxide. The sulphur dioxide is obtained by either burning (imported) elemental sulphur in air:

$$S + O_2 \rightarrow SO_2$$

or by using the sulphur dioxide (by-product) produced when some metals are extracted from their sulphide ores, e.g.,

$$2ZnS + 3O_2 \rightarrow 2ZnO + 2SO_2$$

$$2PbS + 3O_2 \rightarrow 2PbO + 2SO_2$$

The purified sulphur dioxide is then oxidized; the conditions for this equilibrium reaction are given in Chapter 7 (p. 63):

$$2SO_2 + O_2 \rightleftharpoons 2SO_3$$

Although, theoretically, high pressure would favour a higher yield of sulphur(VI) oxide (sulphur trioxide), the additional cost of building high pressure plant would not be justified by the small extra yield of sulphur(VI) oxide which higher pressures produce. For similar economic reasons, vanadium(V) oxide (vanadium pentoxide) is increasingly used because it is cheaper than the alternative catalyst, platinum (also, platinum easily becomes 'poisoned'). Since the reaction between sulphur(VI) oxide and water is extremely vigorous, the gas is absorbed into concentrated sulphuric acid instead, and then later diluted with water as required:

$$SO_3 + H_2SO_4 \rightarrow H_2S_2O_7$$

$$H_2S_2O_7 + H_2O \rightarrow 2H_2SO_4$$

Hot concentrated sulphuric acid is a strong oxidizing agent, e.g., it oxidizes carbon to carbon dioxide (*see* p. 103) and copper to copper(II) ions (*see* before). It has a strong affinity for water, and is therefore used for drying gases with which it does not react. It will also remove combined water, e.g.,

$$\underset{\text{(sugar)}}{C_{12}H_{22}O_{11}} \xrightarrow[(-11H_2O)]{\text{conc. } H_2SO_4} \underset{\text{(carbon)}}{12C}$$

$$\underset{\text{(ethanol)}}{C_2H_5OH} \xrightarrow[(-H_2O)]{\text{conc. } H_2SO_4} \underset{\text{(ethene)}}{C_2H_4}$$

Its reactions with metal halides are discussed under Group VII, p. 123.

$$H-O-\overset{\overset{\displaystyle O}{\|}}{\underset{\underset{\displaystyle O}{\|}}{S}}-O-O-\overset{\overset{\displaystyle O}{\|}}{\underset{\underset{\displaystyle O}{\|}}{S}}-O-H$$

(17)

If moderately concentrated sulphuric acid is electrolysed at 0 °C, peroxodisulphuric (perdisulphuric) acid, $H_2S_2O_8$, is produced, having the structure (17). This acid is a strong oxidizing agent, oxidizing Mn^{2+} ions to manganese(IV) oxide,

$$Mn^{2+} + S_2O_8^{2-} + 2H_2O \rightarrow MnO_2\downarrow + 2SO_4^{2-} + 4H^+$$

and to Mn(VII) using a trace of Ag^+ as catalyst:

$$2Mn^{2+} + 5S_2O_8^{2-} + 8H_2O \rightarrow 2MnO_4^- + 16H^+ + 10SO_4^{2-}$$

Uses of sulphuric acid include: production of 'superphosphate' and ammonium sulphate (fertilizers); manufacture of artificial silk; cleaning of metals (removal of oxide layers) before galvanizing; manufacture of explosives, pigments and dyestuffs; sulphonation of oils to make detergents; accumulators.

12.7 Group VII (F–I)

Some physical properties of the halogens (except astatine) are given in *Table 12.9.* Astatine (not covered in this text) is intensely radioactive, and rare. It does not occur in nature to any appreciable extent, but it has been synthesized by bombarding bismuth with α particles.

Table 12.9 The halogens

Property	F	Cl	Br	I
Electronic configuration	2.7	2.8.7	2.8.18.7	2.8.18.18.7
Physical state (at room temperature)	pale, greenish-yellow gas	Greenish-yellow gas	Dark red, heavy liquid	Grey-black shiny solid
Covalent radius/nm	0.064	0.099	0.111	0.128
Ionic radius/nm	0.133	0.181	0.196	0.219
Electronic affinity/ kJ mol^{-1}	−332.6	−364	−342	−295.4
Ionization energy, I_1/kJ mol^{-1}	1680	1260	1140	1010
Enthalpy (heat) of atomization/kJ mol^{-1}	79.1	121.1	112.0	106.6
Standard electrode potential, $E^{\ominus}$/V, halogen/halide ion	+2.87	+1.36	+1.09	+0.54

The halogens are all one electron short of the noble gas structure, and so they complete their octet by either gaining an electron to form the halide ion, X^-, or by sharing the unpaired *p* electron to form a covalent bond. Ionic halides are formed with the more electropositive metals, whereas covalent compounds are formed with non-metals and weakly electropositive metals (all the halogens exist as diatomic molecules by electron sharing). The covalency maximum increases down the group (since chlorine, bromine and iodine have available *d* orbitals to which electron promotion can occur), e.g., iodine can show a covalency of 7 in IF_7.

As expected from the electronic configurations, the halogens have high electron affinities, the highest value being for chlorine and not fluorine as one might expect. This is because fluorine, being a small molecule, experiences some repulsion between non-bonding pairs of electrons. This results in a relatively low value of dissociation energy and, therefore, enthalpy (heat) of atomization (*see Table 12.9*); when this latter value is fed into a Born–Haber cycle equation, a lower than expected value of electron affinity (for fluorine) is obtained. However, the small size of the F^- ion facilitates its formation in solution (because of its high hydration energy) or in ionic crystals (because of high lattice energies), making fluorine the most electronegative element. Because of this, fluorine causes other elements to exhibit their maximum oxidation states, e.g., SF_6, IF_7, etc; in addition, it also makes fluorine one of the most powerful oxidizing agents (it is not possible to convert the F^- ion into fluorine in solution). The oxidizing power of the halogens decreases down the group. Fluorine displaces chlorine from chlorides, and chlorine displaces bromine and iodine from bromides and iodides (*see* also $E^\ominus$ values: *Table 12.9*), e.g.,

$$Cl_2 + 2I^- \rightarrow 2Cl^- + I_2$$

As expected, electronegativity decreases (or electropositivity increases) down the group; iodine shows some electropositive (or 'metallic') character, and compounds containing the I^+ ion are known. Iodine also shows differences from the other halogens because of its large size, e.g., PI_5 does not exist because there is not room for five iodine atoms around the phosphorus (cf., PF_5, PCl_5, PBr_5). The characteristic test for iodine is that it forms a deep blue colour with starch (and this extremely sensitive test is used in volumetric analysis).

12.7.1 GENERAL PROPERTIES

12.7.1.1 With Water

Fluorine displaces oxygen from water (some hydrogen peroxide and trioxygen (ozone) are formed as well):

$$2H_2O + 2F_2 \rightarrow 4HF + O_2$$

Chlorine is moderately soluble in water, but some of the chlorine reacts to give chloric(I) acid (hypochlorous acid) and hydrochloric acid:

$$Cl_2 + H_2O \rightarrow HCl + HOCl$$

Bromine is moderately soluble in water, whilst iodine is only sparingly soluble in water. To a very small extent the latter reaction occurs for bromine and iodine, i.e., only negligible amounts of HOBr and HOI are formed.

12.7.1.2 With Alkalis

With cold dilute alkali, fluorine gives oxygen difluoride (fluorine monoxide), F_2O, but with warm concentrated alkali, oxygen is formed:

$$2F_2 + 2NaOH \rightarrow 2NaF + F_2O + H_2O$$

$$2F_2 + 4NaOH \rightarrow 4NaF + 2H_2O + O_2$$

With chlorine, bromine and iodine, the following general reaction occurs, to form the halate(I) (hypohalite) ion:

$$X_2 + 2OH^- \rightarrow X^- + XO^- + H_2O$$

but in the case of bromine and iodine, the BrO^- and IO^- ions which are formed rapidly disproportionate, even at room temperature; the ClO^- ion disproportionates rapidly if heated above 75 °C.

$$3ClO^- \rightarrow 2Cl^- + ClO_3^- \text{ [chlorate(V) (chlorate) ion]}$$

$$3BrO^- \rightarrow 2Br^- + BrO_3^- \text{ [bromate(V) (bromate) ion]}$$

$$3IO^- \rightarrow 2I^- + IO_3^- \text{ [iodate(V) (iodate) ion]}$$

12.7.1.3 Properties of the Chlorate(I) (Hypochlorite) Ion

Chloric(I) (hypochlorous) acid, HOCl, is a weak acid and is prepared as described above. It decomposes, on standing, in two possible ways:

$$3ClO^- \rightarrow 2Cl^- + ClO_3^- \text{ (rapid at higher temperatures; } \textit{see} \text{ above)}$$

or

$$2ClO^- \rightarrow 2Cl^- + O_2$$

Addition of hydrochloric acid to a chlorate(I) (hypochlorite) gives chlorine, e.g.,

$$NaOCl + 2HCl \rightarrow NaCl + H_2O + Cl_2$$

Chloric(I) (hypochlorous) acid is a strong oxidizing agent, and in acid solution will oxidize sulphur to sulphuric acid. The reaction between chlorine gas and aqueous bromide or iodide ions may also involve the following reaction:

$$2I^- + \underbrace{HOCl + HCl}_{\text{produced by the water and chlorine}} \rightarrow 2Cl^- + H_2O + I_2$$

In the moist litmus test for chlorine (bleached white) it is the chloric(I) (hypochlorous) acid which causes the bleaching. Consequently, the chlorate(I) (hypochlorite) ion is found in domestic bleaching agents.

12.7.2 HYDROGEN HALIDES

All the hydrogen halides can be prepared by direct combination, the reaction being less vigorous as the group is descended. The reaction with fluorine is explosive; with chlorine it is explosive when irradiated with ultraviolet light (but a jet of hydrogen can be safely burnt in chlorine). With bromine, temperatures of approximately 200 °C and a platinum catalyst are needed; the reaction with iodine is reversible and requires a catalyst and temperatures greater than 200 °C.

At room temperature, hydrogen fluoride is a colourless liquid, the other hydrogen halides being gases (*Table 12.10*). The anomalous boiling point for hydrogen fluoride is because of hydrogen bonding, which causes increased intermolecular attractions (*see* Chapter 3).

Table 12.10 Hydrogen halides

Hydrogen halide	HF	HCl	HBr	HI
B.p./°C	20	−85	−67	−35

Anhydrous hydrogen fluoride is weakly conducting because of the equilibrium

$$2HF \rightleftharpoons H_2F^+ + F^-$$

but the hydrogen halides are predominantly covalent, forming strong acids in aqueous solution, except hydrofluoric acid which is relatively weak. This is mainly because of the greater strength of the H–F bond compared with that of say, H–Cl, and these bonds must be broken in the process of ionization. The H–X bond strengths, in fact, decrease as the group is descended; hence acids become stronger down the group (hydriodic acid is a very strong acid).

$$HX + H_2O \rightarrow H_3O^+ + X^-$$

The measurement of the acid strength of hydrofluoric acid is difficult because of the existence of two equilibria:

$$HF + H_2O \rightleftharpoons H_3O^+ + F^-$$

and

$$HF + F^- \rightleftharpoons HF_2^- \text{ (important in concentrated solution)}$$

Consequently, on dissolving a fluoride salt in aqueous hydrofluoric acid, followed by evaporation, hydrogendifluorides (hydrogen fluorides) are formed:

$$KF + HF \rightleftharpoons KHF_2$$

If the solid KHF_2 is heated, the reaction is reversed.

Hydrofluoric acid is kept in poly(ethene) (Polythene) bottles, and not glass, since it dissolves silicon(IV) oxide (silica) and silicates to form the hexafluorosilicate(IV) ion:

$$SiO_2 + 6HF \rightarrow H_2SiF_6 + 2H_2O$$

hydrolysed ↘ hydrated $SiO_2\downarrow$

The laboratory preparations of the hydrogen halides differ because the latter have different reducing powers (which increase down the group). Hence, hydrogen fluoride and chloride are prepared by heating calcium fluoride and sodium chloride, respectively, with concentrated sulphuric acid:

$$CaF_2 + H_2SO_4 \rightarrow CaSO_4 + 2HF$$

$$NaCl + H_2SO_4 \rightarrow NaHSO_4 + HCl$$

However, the metal halide + concentrated sulphuric acid method cannot be used for preparing hydrogen bromide or iodide, since any HBr or HI formed reduces the sulphuric acid, e.g.,

$$2HBr + H_2SO_4 \rightarrow Br_2 + SO_2 + 2H_2O$$

Hence, HBr and HI are prepared by the hydrolysis of the corresponding phosphorus trihalide:

$$PBr_3 + 3H_2O \rightarrow H_3PO_3 + 3HBr$$

$$PI_3 + 3H_2O \rightarrow H_3PO_3 + 3HI$$

or by the treatment of bromides or iodides with the non-oxidizing phosphoric(V) acid.

12.7.3 HALIDES

Electropositive metals form ionic halides, whereas non-metals and weakly electropositive metals form covalent halides. Elements in their higher valency states give halides which are mainly covalent, e.g., $SnCl_4$ (cf. $SnCl_2$, which has some ionic character). The degree of covalency in a halide increases as the size of the halogen atom increases (note also that the I^- ion is more easily distorted than the F^- ion, giving rise to more covalent character – Fajans's rules, p. 13).

The most obvious method of halide preparation is by direct combination; this (dry) method is essential if the halide in question is easily hydrolysed. This method also makes elements of variable valency exhibit their maximum valency state because of the high electronegativity of fluorine and chlorine (and to a lesser extent bromine), e.g.,

$$2Fe + 3Cl_2 \rightarrow 2FeCl_3$$

$$Sn + 2Br_2 \rightarrow SnBr_4$$

Halides which are not hydrolysed, because they are either highly ionic or insoluble, can be prepared by dissolving the metal, oxide, hydroxide or carbonate in aqueous halogen acid (followed by evaporation of the resultant solution) in the former case, or by a precipitation reaction in the latter case, e.g.,

$$Fe + 2HCl \rightarrow FeCl_2 + H_2$$

$$MgO + 2HCl \rightarrow MgCl_2 + H_2O$$

$$Pb^{2+} + 2I^- \rightarrow PbI_2$$

Generally speaking, the greater the covalency of the halide, the more susceptible it is to hydrolysis, provided that suitable empty orbitals are available to accept incoming lone pairs of electrons from water molecules (*see* hydrolysis of the Group IV hydrides and halides, for example). Some metal halides having appreciable covalent character also undergo hydrolysis, e.g., $BeCl_2$.

The solubilities of chlorides, bromides and iodides tend to be similar, whilst fluorides often have anomalous solubilities, e.g., with silver nitrate solution, solutions of chlorides, bromides and iodides give a white precipitate of silver chloride, a pale yellow precipitate of silver bromide, and a deeper yellow precipitate of silver iodide, respectively. Silver fluoride, however, is soluble (the F^- ion has a high hydration energy; for factors affecting solubility, *see* Chapter 5). On adding ammonia solution to these silver halide precipitates, they dissolve to differing degrees to give the diamminesilver(I) complex ion, $[Ag(NH_3)_2]^+$; the chloride is readily soluble, the bromide is partially soluble and the iodide is only very slightly soluble.

Silver nitrate solution is, in fact, commonly used to estimate chlorides in volumetric analysis. In this analysis, potassium chromate(VI) (chromate) solution is used as the indicator. If standard silver nitrate solution is run from a burette into a solution containing the indicator and a known amount of chloride, silver chloride is preferentially precipitated. Only when all the chloride has been precipitated (as white silver chloride) does the Ag^+ concentration increase to such an extent that the solubility product (*see* Chapter 8) of the silver chromate(VI) is exceeded. Hence, a brick-red precipitate of silver chromate(VI) is then observed, giving the end-point of the titration (N.B., potassium chromate must be used in neutral solutions since Ag_2CrO_4 is soluble in acids).

$$2Ag^+ + CrO_4^{2-} \rightarrow Ag_2CrO_4\downarrow$$

The halogens also form interhalogen compounds, e.g., BrCl, ICl, ICl_3, IF_7, and polyhalide ions, e.g., I_3^-, ICl_2^-.

12.8 Group 0

Some physical properties of the noble gases are given in *Table 12.11*.

Table 12.11 Noble gases

Element	*Symbol*	*Electronic configuration*	*F.p./°C*	*B.p./°C*	I_1/kJ mol^{-1}
Helium	He	2	−269*	−269	2370
Neon	Ne	2.8	−248	−246	2080
Argon	Ar	2.8.8	−189	−186	1520
Krypton	Kr	2.8.18.8	−157	−153	1350
Xenon	Xe	2.8.18.18.8	−112	−108	1170
Radon†	Rn	2.8.18.32.18.8	−71	−62	1040

*Under pressure. †All isotopes of radon are radioactive

The noble gases were originally called the RARE gases because they are found only in very small amounts in the atmosphere and in the Earth's crust. They were also called the INERT gases because of their lack of chemical reactivity. The gases all exist as atoms, i.e., they are monatomic gases. Changes down the group include: atomic size increases; freezing and boiling points increase (because of increased van der Waals forces); they have a greater tendency to form unstable hydrates (*see* below); they have a greater tendency to ionize (because of increased electron shielding).

The only forces of attraction which operate between these atoms are van der Waals forces, which are very weak (*see* Chapter 3). Since these forces are weaker in helium than in any other element, helium is the most difficult gas to liquefy or solidify (*see Table 12.11*). Because of this, helium most closely approaches the ideal gas model (*see* Chapter 4).

Despite the great stability of their electronic configurations, some compounds involving the noble gases have now been made. The molecule PtF_6 has a very high electron affinity, so much so that it is capable of accepting an electron from xenon to form the ionic compound $Xe^+[PtF_6]^-$, which is stable at ordinary temperatures (the noble gases with the higher atomic numbers are most likely to form such compounds because of their lower ionization energies – *see Table 12.11*). Xenon and radon can also form fluorides, e.g., XeF_2, XeF_4, XeF_6.

The shape of xenon tetrafluoride is based on the octahedron, the Xe and F atoms having a square planar configuration because of the influence of the two remaining lone pairs of electrons (*see Figure 2.3*). It can be prepared by heating a 1:5 mixture of Xe and F_2 in a nickel vessel at 400 °C and approximately 6 atmospheres for several hours.

The larger noble gases can also form hydrates, e.g., $Xe.6H_2O$, the polar water molecule polarizing the noble gas atom resulting in a dipole–dipole force of attraction (note that large atoms are more readily polarized because outer electrons are less under the influence of the nucleus).

13

THE FIRST ROW TRANSITION ELEMENTS (Sc–Zn)

13.1 The Elements

One characteristic property of these elements is that they all have partially filled *d* orbitals either in the free atom or in one or more of their chemically important ions (except zinc; *see* below). Some physical properties of these elements are given in *Tables 13.1* and *13.2.*

Table 13.1 The *d*-block elements, scandium to zinc

Element	*Symbol*	*Atomic no. (Z)*	*Electronic configuration*
Scandium	Sc	21	[Ar]$3d^1 4s^2$
Titanium	Ti	22	[Ar]$3d^2 4s^2$
Vanadium	V	23	[Ar]$3d^3 4s^2$
Chromium	Cr	24	[Ar]$3d^5 4s^1$
Manganese	Mn	25	[Ar]$3d^5 4s^2$
Iron	Fe	26	[Ar]$3d^6 4s^2$
Cobalt	Co	27	[Ar]$3d^7 4s^2$
Nickel	Ni	28	[Ar]$3d^8 4s^2$
Copper	Cu	29	[Ar]$3d^{10} 4s^1$
Zinc	Zn	30	[Ar]$3d^{10} 4s^2$

Table 13.2 Physical properties of the *d*-block elements

	Sc	Ti	V	Cr	Mn	Fe	Co	Ni	Cu	Zn
M.p./°C	1400	1677	1917	1903	1244	1539	1495	1455	1083	420
B.p./°C	2477	3277	3377	2642	2041	2887	2877	2837	2582	908
Density/g cm^{-3}	2.99	4.54	6.11	7.19	7.42	7.86	8.90	8.90	8.94	7.13
Enthalpy (heat) of fusion/kJ mol^{-1}	16.11	15.48	17.57	13.81	14.64	15.36	15.23	17.61	13.05	7.38
I_1/kJ mol^{-1}	630	660	650	650	720	760	760	740	750	910
Atomic (covalent) radius/nm	0.144	0.132	0.122	0.117	0.117	0.116	0.116	0.115	0.135	0.131
$E^\ominus$/V, M^{2+}|M	−2.08*	−1.63	−1.20	−0.56	−1.03	−0.41	−0.28	−0.25	+0.34	−0.76

*Sc^{3+}|Sc

Points to be noted from *Table 13.1* are:

(1) the irregularities in electronic configuration at Cr and Cu, illustrating the stability of the d^5 and d^{10} configurations (*see* also p. 6), and
(2) the partially filled *d* orbitals of Sc to Ni. Although Cu has a $3d^{10}4s^1$ configuration (and so *d* orbitals are completely filled), the Cu^{2+} ion has a

$3d^9 4s^0$ configuration, i.e., *d* orbitals are now partially filled, and so Cu^{2+} is a transition metal ion. Note, however, that both Zn and the Zn^{2+} ion ($3d^{10} 4s^0$) have a full complement of *d* electrons and therefore, strictly, are non-transitional (Sc^{3+} is also non-transitional, $3d^0 4s^0$).

From *Table 13.2*, it can be seen that these elements, with the exception of zinc, all have high melting and boiling points, and high enthalpies (heats) of fusion (these values should be compared with the values for the alkali metals, for example, in *Table 11.1*), indicating strong metallic bonding, i.e., the 3*d* and 4*s* electrons are delocalized in the metallic lattice. This strong metallic bonding also explains the high densities of the transition elements (cf. densities of Na, Mg, K, Ca, which are 0.97, 1.74, 0.86 and 1.55 g cm^{-3} respectively).

All the transition elements are metals. They often form alloys with one another (e.g., brass is an alloy of copper and zinc), the atoms being similar in size (*Table 13.2*). Their properties of variable valency (*see* below) and ability to adsorb gases (*see* p. 84) frequently make them good catalysts, and so are useful in industry, e.g., the Contact process (pp. 63 and 119) where a vanadium(V) oxide catalyst is used; the Haber process (pp. 63 and 114) where a catalyst of finely divided iron is used; the catalytic hydrogenation of oils (p. 155) where a nickel catalyst is employed; some polymerization reactions where Ziegler–Natta type catalysts [e.g., $(C_2H_5)_3Al$ with $TiCl_3$ or $TiCl_4$; *see* p. 192) are used. The catalytic action of many of the transition elements is also important in biological systems.

Another property of these elements is that they can take small non-metallic atoms into their lattices (without much distortion of the lattice) to form INTERSTITIAL compounds. For example, titanium is refined in an atmosphere of argon because the former readily takes up nitrogen (argon is too large an atom to be taken into the lattice); *see* also p. 80. Similarly, iron takes up carbon interstitially to form steel, i.e., the carbon atoms prevent the planes of metal atoms from sliding over one another by 'locking' the structure.

Some further common properties of these elements include:

13.1.1 VARIABLE VALENCIES (OR OXIDATION STATES)

The possible oxidation states for the first row transition elements are given in *Table 13.3*. For Sc to Mn, the highest oxidation state corresponds to the total

Table 13.3 Oxidation states of transition elements

Element	*Possible oxidation states*	*Examples*
Sc	+3	Sc_2O_3
Ti	+2, +3, +4	TiO, $TiCl_3$, $TiCl_4$
V	+2, +3, +4, +5	VO, V_2O_3, VO_2, V_2O_5
Cr	+2, +3, +6	CrO, Cr_2O_3, CrO_3
Mn	+2, +3, +4, +6, +7	MnO, Mn_2O_3, MnO_2, K_2MnO_4, $KMnO_4$
Fe	+2, +3	$FeCl_2$, $FeCl_3$
Co	+2, +3, +4	$[Co(H_2O)_6]^{2+}$, $[Co(NH_3)_6]^{3+}$, $[CoF_6]^{2-}$
Ni	+2, +4	NiO, K_2NiF_6
Cu	+1, +2	Cu_2O, CuO
Zn	+2	ZnO

number of *d* and *s* electrons. After Mn, the higher oxidation states are difficult to obtain. Generally speaking, covalent character increases as the oxidation state increases. As previously mentioned, this property of variable valency makes many of the transition elements good catalysts, since they can form intermediate compounds during the reaction (as in homogeneous catalysis) and can also bond reactant molecules to their surfaces during reaction (as in heterogeneous catalysis); *see* also p. 84.

13.1.2 COLOURED IONS AND PARAMAGNETISM

Transition metal ions, except those with empty or completely filled *d* orbitals, possess unpaired electrons and are therefore paramagnetic (*see* p. 110). Hydrated ions possessing unpaired *d* electrons are also coloured (*see Table 13.4*), since colour is caused, in these ions, by absorption of light resulting in the movement of electrons from one *d* level to another (movement cannot occur if the *d* levels

Table 13.4 Colours of aqueous ions of transition elements

$Sc^{3+}_{(aq)}$	$3d^0$	Colourless	$Fe^{2+}_{(aq)}$	$3d^6$	Green
$Ti^{3+}_{(aq)}$	$3d^1$	Violet	$Co^{2+}_{(aq)}$	$3d^7$	Pink
$V^{3+}_{(aq)}$	$3d^2$	Green	$Ni^{2+}_{(aq)}$	$3d^8$	Green
$Cr^{3+}_{(aq)}$	$3d^3$	Green or violet	$Cu^{2+}_{(aq)}$	$3d^9$	Blue
$Mn^{3+}_{(aq)}$	$3d^4$	Violet	$Zn^{2+}_{(aq)}$	$3d^{10}$	Colourless
$Mn^{2+}_{(aq)}$	$3d^5$	Pink			
$Fe^{3+}_{(aq)}$	$3d^5$	Yellow			

are empty or full). Hence, hydrated copper(II) ions are blue because red light of the requisite frequency is absorbed, and white light minus red gives (peacock) blue. The colour of a particular ion can also change with the type of ligand present (*see* below), but an explanation of this is beyond the scope of this text.

13.1.3 COMPLEX ION FORMATION

Although a formal definition is difficult to give, a COMPLEX is formed when a number of molecules or ions (called LIGANDS) combine with a central atom or ion to form an entity in which the number of ligands exceeds the normal valency of the central atom or ion. For example, if copper(II) sulphate is dissolved in water, the following reaction occurs:

$$\underset{\text{white (anhydrous)}}{CuSO_4} + H_2O\ (\text{excess}) \rightarrow \underset{\text{blue hexaaquacopper(II) ion}}{[Cu(H_2O)_6]^{2+}} + SO^{2-}_{4\,(aq)}$$

In this example, the ligands (in this case, water molecules) each donate a pair of electrons to the Cu^{2+} ion, forming co-ordinate bonds. Because water is a donor molecule (Lewis base), transition metal ions in aqueous solution usually form the hexaaqua ion, $[M(H_2O)_6]^{n+}$. However, if other ligands are added, ligand replacement or substitution can occur, e.g.,

$$[Co(H_2O)_6]^{2+} + 4Cl^- \rightleftharpoons [CoCl_4]^{2-} + 6H_2O$$

pink hexaaquacobalt(II) ion; blue tetrachlorocobaltate(II) ion

$$[Fe(H_2O)_6]^{2+} + 6CN^- \rightleftharpoons [Fe(CN)_6]^{4-} + 6H_2O$$

green hexaaquairon(II) ion; yellow-brown hexacyanoferrate(II) ion (ferrocyanide ion)

Several points should be noted:

(1) As stated above, ligand substitution, as in the iron(II) example, can change the colour of the ion.
(2) In the above example, the stable complex $[Fe(CN)_6]^{4-}$ shows none of the properties of its constituent metal ion or ligands, i.e., Fe^{2+} or CN^-.
(3) In naming the complex, negatively charged ligands end in -o, e.g., bromo (Br^-), chloro (Cl^-), cyano (CN^-), hydroxo (OH^-), nitro (NO_2^-), etc; neutral ligands such as ammonia and water are called 'ammine' and 'aqua' respectively. If the complex is in an anionic form, e.g., $[CoCl_4]^{2-}$, then the ending 'ate' is attached to the name of the metal (N.B., some metals take the Latin name when in anionic form, e.g., tin = stannate; iron = ferrate; lead = plumbate; copper = cuprate; silver = argentate).
(4) Water molecules in a complex can be replaced by ligands more willing to donate a lone pair, e.g., NH_3, Cl^-, CN^-, etc. (N.B., Concentration effects are important here.)

When a paramagnetic ion forms a complex, its paramagnetism may diminish or disappear completely, e.g., the Co^{3+} ion (2.8.14., a $3d^6$ system) is paramagnetic, but on forming the hexaamminecobalt(III) complex, $[Co(NH_3)_6]^{3+}$ (2.8.18.8, the 12 additional electrons being supplied by six ammonia molecules), the paramagnetism disappears since there are no longer any unpaired electrons.

The shape of a particular complex depends on the CO-ORDINATION NUMBER of the central atom or ion, i.e., the number of ligands attached to it. For example, co-ordination number 2 occurs with complexes such as $[Cu(NH_3)_2]^+$ (*see* also the $[Ag(NH_3)_2]^+$ complex; Group VII), and the shape is linear. With co-ordination number 4, the tetrahedral configuration is usually observed, e.g., $[FeCl_4]^{2-}$, $[CoCl_4]^{2-}$, $Ni(CO)_4$, etc., with bond angles 109 degrees 28 minutes (this shape is preferred because of electron-pair repulsion theory: *see* Chapter 2). However, square planar configurations also exist particularly in nickel(II) chemistry, e.g., $[Ni(CN)_4]^{2-}$. Co-ordination number 6 is very common, the octahedral configuration being observed, e.g., $[Cu(H_2O)_6]^{2+}$, $[Cr(NH_3)_6]^{3+}$, $[Fe(CN)_6]^{3-}$, etc. (*see* also *Figure 2.3*).

13.2 Stability Constants

On adding ammonia to the Cu^{2+} ion in aqueous solution, the following substitution reactions occur (each stage having its own equilibrium constant):

$$[Cu(H_2O)_6]^{2+} + NH_3 \rightleftharpoons [Cu(H_2O)_5(NH_3)]^{2+} + H_2O$$

$$K_1 = \frac{[[Cu(H_2O)_5(NH_3)]^{2+}]}{[[Cu(H_2O)_6]^{2+}][NH_3]} = 1.8 \times 10^4$$

$$[Cu(H_2O)_5(NH_3)]^{2+} + NH_3 \rightleftharpoons [Cu(H_2O)_4(NH_3)_2]^{2+} + H_2O$$

$$K_2 = \frac{[[Cu(H_2O)_4(NH_3)_2]^{2+}]}{[[Cu(H_2O)_5(NH_3)]^{2+}][NH_3]} = 4.1 \times 10^3$$

$$[Cu(H_2O)_4(NH_3)_2]^{2+} + NH_3 \rightleftharpoons [Cu(H_2O)_3(NH_3)_3]^{2+} + H_2O$$

$$K_3 = \frac{[[Cu(H_2O)_3(NH_3)_3]^{2+}]}{[[Cu(H_2O)_4(NH_3)_2]^{2+}][NH_3]} = 9.6 \times 10^2$$

$$[Cu(H_2O)_3(NH_3)_3]^{2+} + NH_3 \rightleftharpoons [Cu(H_2O)_2(NH_3)_4]^{2+} + H_2O$$

$$K_4 = \frac{[[Cu(H_2O)_2(NH_3)_4]^{2+}]}{[[Cu(H_2O)_3(NH_3)_3]^{2+}][NH_3]} = 1.7 \times 10^2$$

Note that in these reactions, water molecules involved are not included in the equations for the constants K_1 to K_4. These stepwise constants K_1 to K_4 are called STABILITY CONSTANTS. For the overall reaction,

$$[Cu(H_2O)_6]^{2+} + 4NH_3 \rightleftharpoons [Cu(H_2O)_2(NH_3)_4]^{2+} + 4H_2O$$

the OVERALL STABILITY CONSTANT, K (often given the symbol β) is given by:

$$K = K_1 \times K_2 \times K_3 \times K_4 = 1.2 \times 10^{13}$$

Because of the size of these constants, they are usually quoted as logarithms, i.e., $\log_{10}K = 13.08$. The greater the size of $\log_{10}K$ the more stable is the complex. For example, for the formation of the $[CuCl_4]^{2-}$ complex, $\log_{10}K = 5.62$, showing that it is not as stable as the ammine complex above.

13.3 Titanium

Titanium is an abundant element in the Earth's crust, the main ores being ilmenite, $FeTiO_3$, and rutile, TiO_2. Its extraction is described on p. 80 (normal methods of extraction using carbon are not used because of carbide formation; also, the metal is reactive towards oxygen and nitrogen at elevated temperatures). It has a high strength/weight ratio, and is used in the aircraft industry. Titanium combines with most non-metals at high temperatures (its carbide and nitride are interstitial compounds). It is not attacked by mineral acids at normal temperatures, nor by hot aqueous alkali. It does dissolve in hot hydrochloric acid, however, to form the Ti^{3+} salt, and reacts with hot nitric acid to give a hydrated oxide.

Titanium(IV) chloride is prepared by passing chlorine over hot titanium or over a hot mixture of titanium(IV) oxide and carbon:

$$Ti + 2Cl_2 \rightarrow TiCl_4$$

It is a covalent liquid, which fumes in moist air, being hydrolysed by water:

$$TiCl_4 + 2H_2O \rightarrow TiO_2 + 4HCl$$

In concentrated hydrochloric acid, $TiCl_4$ forms the hexachlorotitanate(IV) ion, $[TiCl_6]^{2-}$.

Titanium(IV) oxide (titanium dioxide) is a white solid and, in fact, is used as a white pigment in paints and as a filler in paper. It is amphoteric.

On reducing aqueous titanium(IV) compounds with zinc in acid, solutions containing the $[Ti(H_2O)_6]^{3+}$ ion are obtained; the reduction can be seen by the colour change:

$$\text{Ti(IV) (colourless, } 3d^0) \rightarrow \text{Ti(III) (violet, } 3d^1)$$

The Ti^{3+} ion is easily oxidized by air; it is a fairly strong reducing agent, titanium(III) chloride being used in volumetric analysis.

Titanium(II) chloride can be obtained by the reduction of titanium(IV) chloride with titanium

$$TiCl_4 + Ti \rightarrow 2TiCl_2$$

or by the disproportionation of titanium(III) chloride:

$$2TiCl_3 \rightarrow TiCl_2 + TiCl_4$$

Titanium(II) compounds are few in number and have no aqueous chemistry because of their oxidation by water.

13.4 Vanadium

The important vanadium-containing minerals are vanadinite, $Pb_5(VO_4)_3Cl$, and carnotite, $K(UO_2)VO_4.\frac{3}{2}H_2O$. Its chief use is in alloy steels. It resembles titanium in being corrosion-resistant; it is not attacked by air, water, alkalis or non-oxidizing acids at room temperature, but it does dissolve in nitric and sulphuric acid. At higher temperatures vanadium combines with most non-metals; with oxygen it gives vanadium(V) oxide, V_2O_5 (together with small amounts of lower oxides), and with nitrogen and carbon it gives an interstitial nitride and carbide.

Vanadium(V) oxide, V_2O_5 (prepared as above) is amphoteric; it dissolves in bases to give the vanadate(V) (orthovanadate) ion:

$$V_2O_5 + 6OH^- \rightarrow 2VO_4^{3-} + 3H_2O$$

and also in acids. The V(V) species formed are fairly strong oxidizing agents, e.g., chlorine is evolved when V_2O_5 is dissolved in hydrochloric acid, producing V(IV).

Vanadium(IV) oxide, VO_2, is also amphoteric; it can be prepared as a dark blue solid by the mild reduction of V_2O_5. Vanadium(IV) chloride, prepared by heating vanadium and chlorine, is an oily liquid which is violently hydrolysed by water.

Vanadium(III) oxide, V_2O_3, is a black solid prepared by the reduction of V_2O_5 with fairly powerful reducing agents (e.g., hydrogen or carbon monoxide). It is basic in nature, dissolving in acids to give the green hexaaquavanadium(III) ion, $[V(H_2O)_6]^{3+}$.

Oxidation state (II) is the least stable oxidation state of vanadium. The oxide, VO, is black, having a good electrical conductivity. It is basic, dissolving in acids to give V(II) solutions. There are not many V(II) salts; a complex cyanide, $K_4[V(CN)_6].3H_2O$ is known.

13.5 Chromium

Chromium is a hard, bluish-white metal. It does not tarnish in air, and is therefore used for plating iron and steel, giving them a lustrous appearance. It decomposes steam at red heat to form chromium(III) oxide:

$$2Cr + 3H_2O \rightarrow Cr_2O_3 + 3H_2$$

It dissolves slowly in dilute acids, giving chromium(II) salts, e.g.,

$$Cr + 2HCl \rightarrow CrCl_2 + H_2$$

The important valencies of chromium are 2, 3 and 6; some details are given in *Table 13.5.*

Table 13.5 Valencies (oxidation numbers) of chromium

Valency	*Cation*	*Anion*	*Oxide*	*Properties*
2	Cr^{2+}	–	CrO	Cation readily oxidized, oxide is basic
3	Cr^{3+}	$[Cr(OH)_6]^{3-}$	Cr_2O_3	Cation and anion stable, oxide is amphoteric
6	–	CrO_4^{2-} $Cr_2O_7^{2-}$	CrO_3	Anions stable, oxide is acidic

The most stable and common compounds of chromium are those of chromium(III). On dissolving chromium(III) (chromic) compounds in water, the $[Cr(H_2O)_6]^{3+}$ complex ion is formed; the resulting solution is acidic because of salt hydrolysis (*see* also Group III, p. 100), i.e., the high charge density of the Cr^{3+} ion causes proton release from co-ordinated water molecules:

$$[Cr(H_2O)_6]^{3+} + H_2O \rightleftharpoons [Cr(H_2O)_5(OH)]^{2+} + H_3O^+$$

Hence, if a carbonate solution or a weak base is added, the equilibrium is displaced to the right or forwards (because of proton removal) to form a green-grey precipitate of hydrated chromium(III) hydroxide (cf. aluminium hydroxide, p. 100). If a strong base is used, e.g., sodium hydroxide solution, the precipitate dissolves in excess base to form the chromate(III) (chromite) ion, $[Cr(OH)_6]^{3-}$. If the weak base used is ammonia solution, the precipitated hydroxide dissolves slightly on standing in excess ammonia to give a violet solution of the hexaamminechromium(III) complex ion, $[Cr(NH_3)_6]^{3+}$.

If hydrogen peroxide solution is warmed with a solution of a chromate(III) compound, oxidation to chromium(VI) occurs:

$$2[Cr(OH)_6]^{3-} + 3H_2O_2 \rightarrow \underset{\text{yellow chromate(VI)}}{2CrO_4^{2-}} + 8H_2O + 2OH^-$$

Although the chromate(VI) ion is yellow, some insoluble chromate(VI) compounds are red (e.g., Ag_2CrO_4). Alkaline conditions are needed for the above oxidation since in acid conditions the chromate(VI) ion is converted into the orange dichromate(VI) ion, $Cr_2O_7^{2-}$:

$$\underset{\text{yellow}}{2CrO_4^{2-}} + 2H^+ \underset{\text{Alkali, and certain metal ions}}{\overset{\text{Acid}}{\rightleftharpoons}} \underset{\text{orange}}{Cr_2O_7^{2-}} + H_2O$$

If the metal ion of an insoluble chromate(VI) is added to a dichromate(VI) ion solution, the chromate(VI) is precipitated, e.g.,

$$2Pb^{2+} + Cr_2O_7^{2-} + H_2O \rightarrow \underset{\text{yellow}}{2PbCrO_4\downarrow} + 2H^+$$

If concentrated sulphuric acid is carefully added to a concentrated solution of a chromate(VI) or dichromate(VI), bright red crystals of chromium(VI) oxide (chromium trioxide) are formed. It is a very strong oxidizing agent, and is very soluble in water, forming a solution known as 'chromic acid'. This is used for clearning glassware.

The dichromate(VI) ion in acid solution is a powerful oxidizing agent:

$$Cr_2O_7^{2-} + 14H^+ + 6e^- \rightarrow 2Cr^{3+} + 7H_2O$$

It is used for oxidizing many organic compounds (*see* Chapter 15) and also in volumetric analysis. In the latter situation, a standard dichromate(VI) solution is run from a burette into an acidic solution of the substance undergoing oxidation. Sulphuric acid is usually used for this, but hydrochloric acid can be used instead since the chloride ion is not easily oxidized by the dichromate(VI) [cf. manganate(VII) ion; *see* below]. Since the end point is difficult to detect (as the orange colour of the $Cr_2O_7^{2-}$ ion is replaced by the green colour of the hydrated Cr^{3+} ion), diphenylamine indicator is used which turns from colourless to blue at the end point.

The dichromate(VI) ion will oxidize iron(II) to iron(III), sulphites to sulphates (hence the characteristic test for sulphur dioxide, which turns damp orange dichromate paper green), and iodide ions to iodine.

If a dichromate(VI) solution is reduced by zinc and hydrochloric acid, solutions of Cr(II) are formed:

$$\underset{\text{orange}}{Cr_2O_7^{2-}} \rightarrow \underset{\text{hydrated-green}}{Cr^{3+}} \rightarrow \underset{\text{hydrated-blue}}{Cr^{2+}}$$

13.6 Manganese

Manganese is a hard, grey metal. It is not easily attacked by air, but dissolves readily in dilute acids to form manganese(II) salts, and is rapidly attacked by hot water. It is converted into manganese(II) (manganous) chloride by chlorine. Some details of its important valencies are given in *Table 13.6*. Pure manganese is not very useful as such because of attack by water, but it is used in the manufacture of some steels.

Table 13.6 Valencies (oxidation numbers) of manganese

Valency	*Cation*	*Anion*	*Oxide*	*Properties*
2	Mn^{2+}	–	MnO	Cation stable, oxide is basic
3	Mn^{3+}	–	Mn_2O_3	Cation stable only in complexes, oxide is basic
4	Mn^{4+}	MnO_3^{2-} *	MnO_2	Cation known only in complexes, anion unstable, oxide is amphoteric
6	–	MnO_4^{2-}	–	Anion readily disproportionates to Mn(IV) and Mn(VII)
7	(MnO_3^+)	MnO_4^-	Mn_2O_7	Anion stable, oxide is acidic

*Doubtful composition

The interconversions between these valencies can be illustrated as follows:

$$\underset{\text{manganese(II) ion}}{Mn^{2+}_{(aq)}} \rightleftharpoons \underset{\text{manganese(IV) oxide}}{Mn^{IV}O_2} \rightleftharpoons \underset{\text{manganate(VI) ion}}{Mn^{VI}O_4^{2-}} \rightleftharpoons \underset{\text{manganate(VII) (permanganate) ion}}{Mn^{VII}O_4^-}$$

The most stable state of manganese is manganese(II) (a $3d^5$ system) – the hydrated Mn^{2+} ion is pale pink. Manganese(II) can be oxidized to manganese(IV) using a peroxodisulphate(VI) (perdisulphate) (*see* also Group VI, p. 120) or a chlorate(I) (hypochlorite), e.g.,

$$Mn^{2+} + S_2O_8^{2-} + 2H_2O \rightarrow MnO_2\downarrow + 2SO_4^{2-} + 4H^+$$

Manganese(IV) oxide (manganese dioxide) is a dark-brown solid, insoluble in water. It is reduced by hot concentrated hydrochloric acid to Mn^{2+} (and this reaction can be used as a laboratory preparation of chlorine):

$$MnO_2 + 4HCl \rightarrow MnCl_2 + 2H_2O + Cl_2$$

Manganese(IV) oxide can be oxidized to a manganate(VI) by fusing with an oxidizing agent such as potassium chlorate(V) in the presence of potassium hydroxide:

$$3MnO_2 + 6OH^- + ClO_3^- \rightarrow \underset{\text{green manganate(VI) ion}}{3MnO_4^{2-}} + 3H_2O + Cl^-$$

The manganate(VI) ion is only stable under alkaline conditions, and in neutral or acid conditions disproportionation occurs to give manganese(IV) oxide and the purple manganate(VII) (permanganate) ion:

$$3MnO_4^{2-} + 2H_2O \rightarrow 2MnO_4^- + MnO_2\downarrow + 4OH^-$$

i.e.,

$$Mn(VI) \rightarrow Mn(VII) + Mn(IV)$$

Hence, this method can be used to prepare the manganate(VII) ion. An alternative method of obtaining MnO_4^- from MnO_4^{2-} is to oxidize with chlorine; this avoids the formation of MnO_2:

$$2MnO_4^{2-} + Cl_2 \rightarrow 2MnO_4^- + 2Cl^-$$

The manganate(VII) can be converted into manganate(VI) by heating the former:

$$2KMnO_4 \rightarrow K_2MnO_4 + MnO_2 + O_2$$

Potassium manganate(VII) is a powerful and useful oxidizing agent, both in acid and alkaline conditions:

$$MnO_4^- + 8H^+ + 5e^- \rightarrow Mn^{2+} + 4H_2O$$

$$MnO_4^- + 2H_2O + 3e^- \rightarrow MnO_2 + 4OH^-$$

However, acidic conditions are normally used to avoid the formation of insoluble manganese(IV) oxide. The acid used is normally sulphuric acid, since the MnO_4^- ion can oxidize the Cl^- ion of hydrochloric acid to chlorine [cf. dichromate(VI); *see* before].

$$2KMnO_4 + 16HCl \rightarrow 2KCl + 2MnCl_2 + 8H_2O + 5Cl_2$$

If the potassium manganate(VII) is being used in volumetric work, nitric acid cannot be used either since the latter is an oxidizing agent and can interfere with the oxidation process. In volumetric work, the colour change at the end point is virtually purple to colourless, because the hydrated Mn^{2+} ion concentration is too low to colour the solution pink. Examples of substances oxidized by the manganate(VII) ion include: iron(II) to iron(III); hot ethanedioates (oxalates) to carbon dioxide; sulphites to sulphates; nitrites to nitrates. The manganate(VII) ion is also extremely useful in the oxidation of organic compounds, e.g., oxidation of alcohols (*see* Chapter 15).

13.7 Iron

Iron is the second most abundant metal in the Earth's crust (aluminium being the most abundant); its ores include: haematite, Fe_2O_3, limonite, $Fe_2O_3.H_2O$, magnetite Fe_3O_4, siderite, $FeCO_3$, and iron pyrites, FeS_2.

13.7.1 EXTRACTION

The extraction consists of two processes; first, the iron is obtained from its oxide (called SMELTING) in the blast furnace, and secondly the metal is REFINED. Smelting is carried out by feeding iron ore, coke and limestone, in alternate layers, in through the top of the blast furnace, whilst a blast of hot air is forced in at the bottom. Hence, combustion of coke to carbon monoxide occurs:

$$2C + O_2 \rightarrow 2CO$$

The temperature of the furnace ranges from about 1600 °C at the bottom to about 200 °C at the top. Near the top of the furnace, where it is cooler, carbon monoxide is the main reducing agent:

$$3CO + Fe_2O_3 \rightleftharpoons 2Fe + 3CO_2$$

and

$$CO + Fe_2O_3 \rightarrow 2FeO + CO_2$$

The iron(II) oxide (ferrous oxide) produced is reduced by coke further down the furnace where it is hotter:

$$FeO + C \rightarrow Fe + CO$$

The gases escaping from the top of the furnace contain up to 28 per cent carbon monoxide and this is burnt to heat the incoming air at the bottom. The limestone in the furnace decomposes to form calcium oxide, which reacts with silicon(IV) oxide (silica) impurity to form a slag; the carbon dioxide also produced is reduced to the monoxide by the coke:

$$CaCO_3 \rightarrow CaO + CO_2 \quad (\text{and } CO_2 + C \rightarrow 2CO)$$

then

$$CaO + SiO_2 \rightarrow CaSiO_3$$

This slag floats on top of the molten iron, hence preventing it from being re-oxidized. The iron produced, which is run off at the bottom of the furnace into moulds, is called PIG IRON. It contains between 5 and 10 per cent impurities (mainly carbon, but some phosphorus, silicon and manganese).

The iron is then refined to remove impurities, followed by the addition of specific elements depending on the typè of steel required. In the Bessemer (refining) process, molten pig iron is poured into a steel vessel lined with silica; a blast of air is blown through the molten iron and the impurities burn off. This process is never used if the phosphorus content is above 0.1 per cent; in the latter situation, phosphorus can be removed by using a basic lining of calcined dolomite ($MgCO_3.CaCO_3$) instead. The basic oxides (MgO and CaO) remove the phosphorus as phosphates, forming a slag which can be sold as a fertilizer. In both processes, the carbon content of the steel is brought to the required level by adding an Fe–C–Mn alloy.

All steels have a carbon content of 0.1–1.5 per cent. A steel having a carbon content of 0.1 per cent is called mild steel, whereas high carbon steel contains up to 1.5 per cent. Addition of other elements imparts specific properties on the steel, e.g., addition of manganese gives the steel elasticity and high tensile strength; stainless steel, having high resistance to chemical attack, contains up to 15 per cent chromium or up to 18 per cent chromium and nickel; steel containing tungsten and vanadium has great hardness and is used for making high speed drills.

The process of ore reduction in a blast furnace can be treated in terms of free energy. In many chemical reactions, there is an increase in disorder (ΔS is positive) and an evolution of heat (ΔH is negative); *see* Chapter 5. Both ΔH and ΔS are related by the expression

$$\Delta G = \Delta H - T\Delta S$$

where ΔG is the change in FREE ENERGY. Therefore, if ΔH is negative and ΔS

is positive in a reaction, then ΔG takes a negative value. Consider the following reactions:

(1) $Fe + \frac{1}{2}O_2 \rightarrow FeO$; $\Delta G = -244$ kJ mol^{-1} (25 °C)

(2) $C + \frac{1}{2}O_2 \rightarrow CO$; $\Delta G = -137$ kJ mol^{-1} (25 °C)

Reversing equation (1) means that +244 kJ mol^{-1} of free energy are required to produce iron from iron(II) oxide, which is more than can be supplied if carbon is used as the reducing agent. However, free energy changes depend on temperature; a plot of ΔG *versus* temperature constitutes an ELLINGHAM DIAGRAM. On consulting such diagrams, it is seen that the reduction reaction using carbon now becomes feasible at temperatures above 1000 °C, since sufficient free energy can be supplied from process (2) for the iron(II) oxide to be reduced.

13.7.2 THE RUSTING OF IRON

Iron is slowly converted into hydrated iron(III) oxide (or rust) by oxygen and moisture (both are essential). It is an electrolytic process and so is accelerated by carbon dioxide (which produces carbonic acid) and the presence of salt (used on roads in winter) since they form electrolytes. Rust can be prevented by coating the iron with a metal having a more negative electrode potential than itself (*see* p. 78).

13.7.3 GENERAL PROPERTIES

Iron dissolves in dilute acids forming iron(II) salts and hydrogen:

$$Fe + 2H^+ \rightarrow Fe^{2+} + H_2$$

With dilute nitric acid, reduction to ammonia occurs:

$$4Fe + 10H^+ + NO_3^- \rightarrow 4Fe^{2+} + NH_4^+ + 3H_2O$$

(i.e., the ammonia is converted into ammonium nitrate).

Heating iron with concentrated sulphuric acid reduces some of the acid to sulphur dioxide:

$$Fe + 2H_2SO_4 \rightarrow FeSO_4 + 2H_2O + SO_2$$

Concentrated nitric acid, however, renders iron 'passive' because of the formation of an impervious film of iron(II) diiron(III) oxide (ferrosoferric oxide), Fe_3O_4.

The two common oxidation states of iron are +2 and +3; the +3 state is the most stable because it is a $3d^5$ system [Fe(II) is $3d^6$]. Both the Fe^{2+} and Fe^{3+} ions form hexaaqua complexes when in aqueous solution (for colours, *see Table 13.4*). Soluble iron(III) salts give acidic solutions because of hydrolysis (*see* also chromium):

$$[Fe(H_2O)_6]^{3+} + H_2O \rightleftharpoons [Fe(H_2O)_5(OH)]^{2+} + H_3O^+$$

The Fe^{2+} ion is fairly stable in acid solution, but is easily oxidized by such oxidizing agents as potassium manganate(VII) (permanganate), potassium dichromate(VI), nitric acid, etc.

Example:

$$5Fe^{2+} + MnO_4^- + 8H^+ \rightarrow 5Fe^{3+} + Mn^{2+} + 4H_2O$$

In neutral or alkaline solution, iron(II) is easily oxidized (even by atmospheric oxygen) to iron(III), i.e., if sodium hydroxide solution is added to an iron(II) salt in aqueous solution, green iron(II) hydroxide is precipitated. This precipitate is in fact white; the green colour is due to an intermediate stage in its oxidation to the brown coloured iron(III) hydroxide.

$$Fe^{2+}_{(aq)} + 2OH^- \rightarrow Fe(OH)_2\downarrow \text{ (green)}$$

Iron(III) hydroxide can be prepared by adding an alkali or a carbonate (proton removal drives above equilibrium forwards) to an iron(III) salt in aqueous solution:

$$Fe^{3+}_{(aq)} + 3OH^- \rightarrow Fe(OH)_3\downarrow \text{ (brown)}$$

[N.B., both iron(II) and iron(III) hydroxides can be formed by adding ammonia solution to their respective aqueous salts, but in both cases no hexaammine complex is formed, even in excess ammonia.]

On adding potassium cyanide solution to aqueous Fe^{2+} ions, a yellow solution containing the hexacyanoferrate(II) (ferrocyanide) ion is formed [potassium hexacyanoferrate(II) is a pale yellow crystalline solid]:

$$[Fe(H_2O)_6]^{2+} + 6CN^- \rightleftharpoons [Fe(CN)_6]^{4-} + 6H_2O$$

It is a stable complex (not giving the reactions of Fe^{2+} or CN^-), and shows how a change of ligand can confer stability on the Fe^{2+} ion. The hexacyanoferrate(II) complex ion can be oxidized to the hexacyanoferrate(III) (ferricyanide) complex ion by oxidizing agents such as chlorine, and can be separated from solution as dark red crystals:

$$2[Fe(CN)_6]^{4-} + Cl_2 \rightarrow 2[Fe(CN)_6]^{3-} + 2Cl^-$$

One way of distinguishing between Fe^{2+} and Fe^{3+} ions is to add ammonium thiocyanate; no change occurs with Fe^{2+}, but with Fe^{3+} a blood-red colouration is formed.

13.8 Cobalt and Nickel

These elements have many similarities: both are below iron in the electrochemical series (*see* $E^\ominus$ values; *Table 13.2*); neither is attacked by water or air at normal temperatures; both dissolve very slowly in dilute hydrochloric and sulphuric acids but fairly readily in dilute nitric acid (concentrated nitric acid renders them passive). The most stable oxidation state for these elements is +2, although cobalt is stable in the +3 state when in complexes (*see* below). Cobalt(III) is a strong oxidizing agent [itself being reduced to cobalt(II)]. Aqueous solutions of cobalt(II) and nickel(II) salts contain the $[Co(H_2O)_6]^{2+}$ and $[Ni(H_2O)_6]^{2+}$ complex ions respectively (*see Table 13.4* for colours).

Both form hydroxides on adding OH^- ions (from ammonia or sodium hydroxide solutions) to their aqueous salts:

$$Co^{2+}_{(aq)} + 2OH^- \rightarrow Co(OH)_2\downarrow \text{ (blue, but turns pink on standing)}$$

$$Ni^{2+}_{(aq)} + 2OH^- \rightarrow Ni(OH)_2\downarrow \text{ (green)}$$

When using ammonia solution, both hydroxides dissolve in excess ammonia to give hexaammine complexes. In the case of $Ni(OH)_2$ the bluish-violet $[Ni(NH_3)_6]^{2+}$ complex is formed. With $Co(OH)_2$ a pale straw coloured solution forms which, on standing in air, turns from yellow to darker brown (this process can be accelerated by bubbling air through the solution or by adding hydrogen peroxide). The latter situation can be represented as follows:

$$Co(OH)_2 \xrightarrow{\text{excess } NH_3} [Co(NH_3)_6]^{2+} \xrightarrow[\text{(rapid)}]{O_2 \text{ or } H_2O_2} [Co(NH_3)_6]^{3+}$$

Hence, the hexaamminecobalt(II) complex is unstable, whilst the hexaamminecobalt(III) complex is stable. One theory is that in the formation of the latter, both the inner incomplete shell and the outer shell (of the central metal ion) accept the incoming electrons from the ammonia ligands, and hence the complex ion achieves the krypton configuration $\{Co^{3+}$, 2.8.14; $[Co(NH_3)_6]^{3+}$, 2.8.18.8$\}$. In the case of the hexaamminecobalt(II) complex, it is thought that complex formation occurs either by ion–dipole attractions, or that the ammonia ligands donate electrons to the outer shell only $\{Co^{2+}$, 2.8.15; $[Co(NH_3)_6]^{2+}$, 2.8.15.12$\}$.

The chlorides of both cobalt and nickel may be obtained in the anhydrous state by heating the respective metals in chlorine; $NiCl_2$ is a yellow solid, and $CoCl_2$ is blue. If concentrated hydrochloric acid is added to a pink aqueous solution of cobalt chloride, the solution turns blue:

$$[Co(H_2O)_6]^{2+} + 4Cl^- \rightleftharpoons [CoCl_4]^{2-} + 6H_2O$$

$[Co(H_2O)_6]^{2+}$	$[CoCl_4]^{2-}$
pink	blue
hexaaquacobalt(II) ion	tetrachlorocobaltate(II) ion
(octahedral)	(tetrahedral)

A characteristic test used for detecting Ni^{2+} ions is to add butanedione dioxime (dimethylglyoxime); a red precipitate is formed.

13.9 Copper

Copper has a low position in the electrochemical series, indicating little chemical reactivity (*see* also $E^{\ominus}$ value; *Table 13.2*). In a dry atmosphere, it is not attacked at normal temperatures, but is oxidized to black copper(II) (cupric) oxide when heated. Hydrogen is not obtained with a dilute acid, because of copper's position in the electrochemical series relative to hydrogen. Copper is not attacked by either dilute hydrochloric or sulphuric acid,* but dissolves in dilute nitric acid to give nitrogen oxide and in concentrated nitric acid to give nitrogen dioxide:

$$3Cu + 8HNO_3 \rightarrow 3Cu(NO_3)_2 + 4H_2O + 2NO$$

$$Cu + 4HNO_3 \rightarrow Cu(NO_3)_2 + 2H_2O + 2NO_2$$

With hot concentrated sulphuric acid, sulphur dioxide is produced:

$$Cu + 2H_2SO_4 \rightarrow CuSO_4 + 2H_2O + SO_2$$

*If hydrogen peroxide is present to promote oxidation, copper dissolves to form copper sulphate in dilute sulphuric acid.

The two common oxidation states of copper are +1 and +2 (+2 being the more stable). Complex formation occurs with both oxidation states. Copper(II) salts in aqueous solution form the blue $[Cu(H_2O)_6]^{2+}$ ion. If, however, copper(II) sulphate is crystallised from solution, the pentahydrate, $CuSO_4.5H_2O$, is formed. Here, each Cu^{2+} ion is surrounded by four water molecules, the fifth being held by hydrogen bonds.

On adding ammonia solution to a solution of a copper(II) salt, a pale blue precipitate of copper(II) hydroxide is produced, which dissolves in excess ammonia to give a deep blue solution containing the diaquatetraamminecopper(II) complex ion, $[Cu(NH_3)_4(H_2O)_2]^{2+}$:

$$Cu^{2+}_{(aq)} + 2OH^- \rightarrow \underset{\text{(pale blue)}}{Cu(OH)_2\downarrow} \xrightarrow{\text{Excess ammonia}} \underset{\text{(deep blue solution)}}{[Cu(NH_3)_4(H_2O)_2]^{2+}}$$

On adding dilute hydrochloric acid to copper(II) hydroxide or carbonate, a blue–green solution of the chloride, $CuCl_2$, is produced; however, further addition of chloride ions (e.g., from concentrated hydrochloric acid) produces the yellow tetrachlorocuprate(II) complex ion, $[CuCl_4]^{2-}$ (it often looks green because of the presence of some blue $[Cu(H_2O)_6]^{2+}$ as well).

If copper(II) chloride is boiled for several minutes with concentrated hydrochloric acid and then copper, the $[CuCl_4]^{2-}$ complex is reduced to the dichlorocuprate(I) complex, $[CuCl_2]^-$:

$$[Cu(H_2O)_6]^{2+} + 4Cl^- \rightleftharpoons [CuCl_4]^{2-} + 6H_2O$$

$$[CuCl_4]^{2-} + Cu \rightarrow 2[CuCl_2]^-$$

On pouring the $[CuCl_2]^-$ complex into a large excess of water, the Cl^- ion concentration is effectively reduced and a white precipitate of copper(I) (cuprous) chloride, CuCl, is formed. On adding ammonia to this precipitate, the soluble (colourless) diamminecopper(I) complex, $[Cu(NH_3)_2]^+$, is formed, which is rapidly oxidized in air to the deep blue $[Cu(NH_3)_4(H_2O)_2]^{2+}$ complex.

Attempts to prepare copper(II) iodide by adding potassium iodide to a copper(II) salt solution fail, since white copper(I) iodide is precipitated instead:

$$2Cu^{2+}_{(aq)} + 4I^- \rightarrow 2CuI\downarrow + I_2$$

[This reaction provides a method of estimating copper(II) since the liberated iodine can be titrated with standard sodium thiosulphate(VI) solution.] Similarly, addition of a cyanide to copper(II) salt solutions gives a white precipitate of copper(I) cyanide, CuCN.

Copper(I) sulphate can be prepared as a white powder by heating dimethyl sulphate with copper(I) oxide:

$$(CH_3)_2SO_4 + Cu_2O \rightarrow Cu_2SO_4 + (CH_3)_2O$$

Copper(I) compounds cannot exist in aqueous solution because of disproportionation; hence copper(I) sulphate, which is soluble in water, will disproportionate in its presence:

$$Cu_2SO_4 \xrightarrow{\text{water}} CuSO_4 + Cu\downarrow$$

(i.e., $2Cu^+ \rightarrow Cu + Cu^{2+}$)

Note, however, that the copper(I) ion can be stabilized in aqueous solution by complex formation, e.g., $[CuCl_2]^-$ (*see* above).

14

INTRODUCTION TO ORGANIC CHEMISTRY

14.1 Organic Chemistry

Organic chemistry is the chemistry of carbon compounds. However, the oxides of carbon, and carbonates, whilst being compounds of carbon, are usually studied in inorganic chemistry. Organic compounds are extremely important because they are not only present in living matter (*see*, for example, Chapter 16) but also in commercial products, e.g., petroleum products, rubber, plastics, dyes, drugs, and so on.

14.2 The Purification of Organic Compounds

During an organic preparation, the desired product may be contaminated with side-products of the reaction or impurities. The following procedures represent some ways in which the desired sample can be purified.

14.2.1 CRYSTALLIZATION

This method, which is used for solids, depends on finding a solvent which readily dissolves the required product, X, when hot but only to a small extent when cold. The crude product, X, is dissolved in the minimum amount of boiling solvent, followed by filtration to remove any insoluble impurities, and is then allowed to cool; the impurities remain in solution, and most of X crystallizes out. The crystals are filtered off, often under reduced pressure using a Buchner funnel and flask. When most of the solvent has been sucked out, the crystals may be dried between filter papers, or in a vacuum oven (the latter can be heated to assist solvent removal provided X is thermally stable at that temperature).

The purity of the crystals may be determined by a melting point determination, since the presence of an impurity not only depresses the melting point (*see* also Chapter 6, p. 52), but makes the compound melt over a range of temperature (in contrast to a pure substance, which has a sharp melting point). If the melting point indicates that X is not pure, the crystallization process can be repeated until it is pure.

14.2.2 SOLVENT EXTRACTION

Sometimes, in an organic preparation, the required compound, X, is produced in aqueous solution. A satisfactory method of extracting X is to shake the aqueous solution with an organic solvent which is immiscible with water but which is a solvent for X. Ethoxyethane (diethyl ether) is commonly used for this process since

(1) it is immiscible with water;
(2) it is relatively inert;
(3) it has a low boiling point and so can easily be distilled off from the compound.

The process is carried out in a separating funnel, in which the aqueous solution and ethoxyethane are shaken together. When equilibrium is obtained (*see* also Chapter 6, p. 55), the lower aqueous layer is run off *via* the tap at the

Table 14.1 Drying agents

Substance	*Uses and limitations*
Calcium chloride	Cannot be used for alcohols, phenols or amines since it combines with them, nor for acidic liquids because it usually contains some calcium hydroxide.
Calcium oxide ('quicklime')	Usually used for alcohols; cannot be used for acidic compounds or esters.
Potassium hydroxide	Suitable for amines; cannot be used for acids, phenols or esters since it combines with them.
Potassium carbonate	Suitable for amines, particularly if a strongly alkaline drying agent is to be avoided.
Sodium sulphate (anhydrous)	Can be used for almost everything, but is slow.
Magnesium sulphate (anhydrous)	Can be used for almost everything, and is faster than sodium sulphate.
Calcium sulphate ('Drierite')	Can be used with all liquids.
Sodium wire	Used particularly for ethers; cannot be used for any compound affected by alkalis, or which may be reduced easily by evolved hydrogen $2Na + 2H_2O \rightarrow 2NaOH + H_2$

bottom of the funnel, and the ethoxyethane layer is then dried (several extractions may be carried out, if necessary, the ethoxyethane extracts being subsequently combined). Distillation to remove the ethoxyethane yields the required compound.

Some drying agents, and their limitations, are given in *Table 14.1.*

14.2.3 DISTILLATION

Purification by distillation can be used when the impurities in a liquid are non-volatile. The impure liquid is placed in the distillation flask together with a few pieces of broken porcelain to prevent 'bumping'. The flask is heated, the liquid boils, and the vapour of the liquid is then condensed and collected in a suitable receiver; the impurities are left in the distillation flask. If, however, the impurities themselves are volatile liquids, then separation can be achieved by using a fractionating column; details of this are given in Chapter 6, p. 56. If the required material decomposes fairly easily on heating or has a very high boiling point, the distillation can be performed under reduced pressure.

14.2.4 STEAM DISTILLATION

Many water-insoluble substances (both liquids and solids) may be purified by distillation in a current of steam, provided that the substance is volatile, and the impurities are non-volatile, under these conditions. Steam distillation yields the purified compound and water as the distillate. The compound may be separated from the water by filtration in the case of a solid, or by extraction with ethoxyethane in the case of a liquid. Further details are given in Chapter 6, p. 57.

14.2.5 CHROMATOGRAPHY

In adsorption column chromatography, a mixture (or impure material) can be split up into its components by allowing them to be selectively adsorbed from solution by a suitable substance such as alumina, Al_2O_3. Experimentally, this can be achieved by allowing a solution of the impure substance to pass slowly down a long column packed with the adsorbent material. The extent to which this adsorbent material adsorbs the various components of the mixture can vary considerably, and hence separation of the mixture into bands can result. If the components are coloured, separation can be easily observed. The bands can be mechanically separated by actually removing the column material in sections or, more conveniently, further solvent can be allowed to run through the column to wash or ELUTE the bands from the column material.

In partition chromatography, the components of the mixture are partitioned (*see* p. 55) between the solvent and the water held by the packing material. If very small quantities of material are being used, separation of a mixture can be achieved by thin layer or paper chromatography. Here, a solution of the mixture passes over a thin layer of alumina, cellulose, etc., and the various components are adsorbed or partitioned at different rates. The ratio of the distance travelled by a particular solute to the distance travelled by the solvent front is known as the R_f value, and is a characteristic of a particular substance.

Chromatography is used for testing the purity of a sample (and for analysis) as well as for separation. Gas–liquid chromatography (or vapour phase chromatography) is particularly useful for the analysis of volatile substances. The material under test (either gas or volatile liquid) is injected into the apparatus and is carried by a stream of gas (usually helium or nitrogen) into a long heated column packed with a porous solid (e.g., crushed fire-brick) impregnated with a non-volatile liquid or oil. Gas–liquid partitioning occurs, the retention time in the column being characteristic of a particular substance. Detection is usually achieved by measuring the changes in thermal conductivity of the effluent gases.

14.3 Empirical, Molecular and Structural Formulae

Once the purity of a certain compound has been established, the next problem is to identify it; this involves determining its empirical formula, then molecular

```
  H         O
   \       //
H---C---C
   /       \
  H         O---H
```

(1)

formula, and finally its structural formula. The EMPIRICAL FORMULA is the simplest formula that shows the ratio of the atoms in a molecule, whilst the MOLECULAR FORMULA gives the number, as well as the ratio, of the atoms in a molecule. The STRUCTURAL FORMULA shows how these atoms are arranged in the molecule, e.g., ethanoic (acetic) acid has the empirical formula CH_2O, molecular formula $C_2H_4O_2$, and the structural formula (1).

14.3.1 DETERMINATION OF EMPIRICAL FORMULA

The first step is detection of the elements (qualitative analysis). Carbon and hydrogen may be detected by heating a small amount of the compound with copper(II) oxide; any carbon present is oxidized to carbon dioxide (detected with lime-water) and any hydrogen present is oxidized to water [detected with anhydrous copper(II) sulphate, which turns blue in the presence of water]. For detecting nitrogen, sulphur and the halogens, the Lassaigne method is used, which involves fusing a small amount of the compound with sodium. This process converts any nitrogen present into sodium cyanide, halogens into sodium halides, and sulphur into sodium sulphide; these sodium salts can then be tested for by standard inorganic tests. In the case of metals, the organic compound is heated in air or a stream of oxygen and the residue examined, again, by standard inorganic tests. A few metals can be detected by their characteristic flame test as shown in *Table 14.2.*

Table 14.2 Colours in flame tests

Element	Potassium	Sodium	Calcium	Barium	Copper
Flame colour	Lilac	Golden yellow	Brick red	Pale green	Bright green

Once it has been determined which elements are present, a quantitative analysis can be carried out. This involves microanalytical techniques, requiring only very small quantities of the material (3–5 mg). From the data obtained, the empirical (simplest) formula can be calculated.

Example: An organic compound, containing only carbon, hydrogen and oxygen, was found to contain 40.0 per cent carbon and 6.7 per cent hydrogen. Calculate its empirical formula (A_r: C = 12, H = 1, O = 16).

Element	C	H	O
Percentage	40.0	6.7	53.3 (by difference)
Atomic ratio	40.0/12 = 3.33	6.7/1 = 6.7	53.3/16 = 3.33
Simplest whole number ratio	1	2	1

Hence, the empirical formula is CH_2O.

14.3.2 DETERMINATION OF MOLECULAR FORMULA

If the relative molecular mass of the compound is now found by one of the standard physical methods (*see* Chapter 6 and also mass spectrometry – Chapter 1), the molecular formula can then be obtained, since the latter must be the same as, or a multiple of, the empirical formula. For example, if, in the previous example, the relative molecular mass of the organic compound was found to be 60, then the molecular formula would be $C_2H_4O_2$. If it was found to be 180, then the molecular formula would be $C_6H_{12}O_6$.

14.3.3 DETERMINATION OF STRUCTURAL FORMULA

Once the molecular formula is known, the structural formula can then be determined. This can be a long and complicated process. A detailed examination of the reactions of the compound can give information as to the functional groups present; measurements of the different types of spectrum are also very valuable in elucidating structure. The following are some of the techniques available.

14.3.3.1 Mass Spectrometry

With this technique (*see* also Chapter 1), not only can values of relative molecular mass be obtained, but also values of the relative molecular masses and relative abundances of molecular fragments produced (when fragmentation occurs). Hence, structural information can be deduced from these data.

14.3.3.2 Infrared Spectroscopy

Atoms joined by a covalent bond are in a state of continual vibration, each type of bond vibrating at its own characteristic frequency, i.e., many bonds or groups give absorptions in characteristic regions of the infrared spectrum. For example, the –OH group: 3600 cm^{-1}; the $–NH_2$ group: 3400 cm^{-1}; the $>C=O$ group: 1750–1600 cm^{-1}, and so on. Hence, various types of bonds and functional groups can be detected.

14.3.3.3 Nuclear Magnetic Resonance Spectroscopy

This technique is based on the fact that the hydrogen nucleus, when placed in a magnetic field, absorbs radiant energy in the radiofrequency region; this absorption is affected by the environment of the nucleus. From the wavelength and intensity of absorption, the numbers and positions of hydrogen atoms in a molecule can be determined.

14.4 Isomerism

The occurrence of two or more compounds with the same molecular formula, but having one or more different physical or chemical properties, is known as ISOMERISM. The two main types of isomerism are

(1) Structural isomerism
(2) Stereoisomerism, which can be further divided into GEOMETRICAL and OPTICAL isomerism.

14.4.1 STRUCTURAL ISOMERISM

Compounds which have the same molecular formula but different structural formulae exhibit structural isomerism. Examples include ethanol (2) and

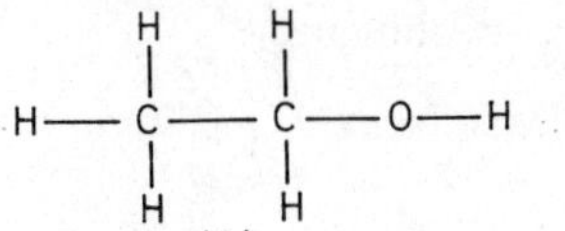

$$H-\underset{H}{\overset{H}{C}}-\underset{H}{\overset{H}{C}}-O-H$$

(2)

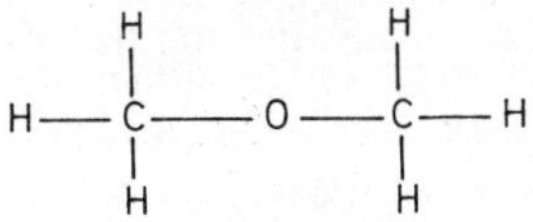

$$H-\underset{H}{\overset{H}{C}}-O-\underset{H}{\overset{H}{C}}-H$$

(3)

$$CH_3-CH_2-CH_2-CH_3$$

(4)

$$CH_3-\underset{CH_3}{CH}-CH_3$$

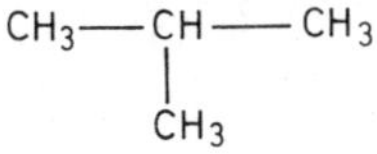

(5)

methoxymethane (dimethyl ether) (3), and butane (4) and 2-methylpropane (5). Structural isomers have different physical and chemical properties.

14.4.2 STEREOISOMERISM

Stereoisomerism is concerned with isomers having the same structure but different spatial arrangements. Two kinds of stereoisomerism are known – geometrical isomerism and optical isomerism.

14.4.2.1 Geometrical Isomerism

Carbon atoms cannot rotate about a double bond (unlike atoms joined by a single bond which can rotate). Therefore, in some compounds containing a double bond, this restricted rotation means that the positions of atoms or groups attached to the double bond are fixed in space. Examples of this occurrence are the two geometrical isomers *cis*- and *trans*-butenedioic acids (maleic and fumaric acids respectively). Geometrical isomers differ in physical

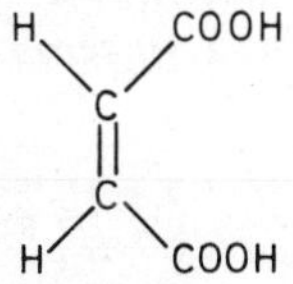

H, COOH
C
||
C
H, COOH

cis-Butenedioic acid

(6)

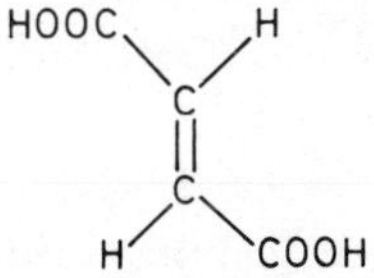

HOOC, H
C
||
C
H, COOH

trans-Butenedioic acid

(7)

properties and many of their chemical properties. For example, the *cis*-isomer has a lower melting point than the *trans*-isomer because the former has two large carboxyl groups on the same side of the double bond and these can interfere with one another, making the compound less stable.

14.4.2.2 Optical Isomerism

If a carbon atom has joined to it four different atoms or groups (in a tetrahedral configuration, therefore), the carbon atom is referred to as an ASYMMETRIC carbon atom, and the compound and its mirror image are non-superimposable, i.e., two different molecules are possible, as shown in structures (8) and (9). The situation is analogous to left and right hand gloves; from whichever point they are examined, they are never the same. Compounds containing

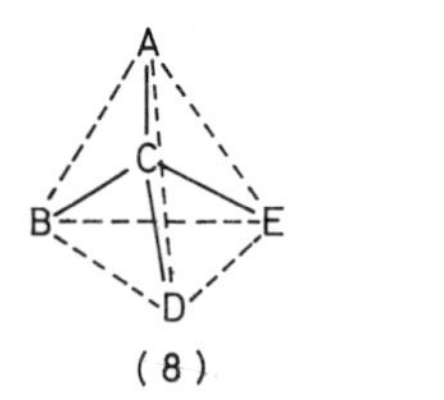

(8)

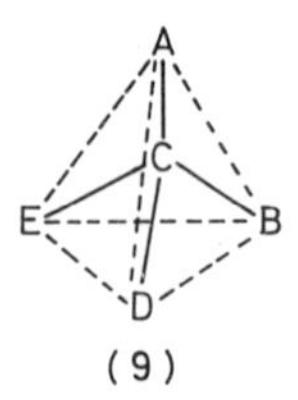

(9)

an asymmetric carbon atom show optical activity, i.e., they rotate the plane of plane polarized light (light which is vibrating in one plane only). One form of the molecule rotates the light to the right [and is called dextrorotatory or the (+) form] whilst the other rotates the light to the left [and is called laevorotatory or the (–) form]. The two optical isomers (or mirror images) are called ENANTIOMERS or ENANTIOMORPHS, and they possess the same physical and chemical properties other than the direction in which they rotate plane polarized light. An equal mixture of the two forms (called a RACEMIC mixture) is optically inactive, however, since the rotation by one form cancels out the other. Examples of optically active compounds include 2-hydroxypropanoic acid (lactic acid) (10) and 2-aminopropanoic acid (alanine) (11); *see* also Chapter 16.

2-Hydroxypropanoic acid
(10)

2-Aminopropanoic acid
(11)

Separation of a racemic mixture (or racemate) into the pure enantiomers is known as RESOLUTION. Pasteur devised the three classical methods of resolution.

(1) By crystallization, and then separating by hand the two types of crystal. This method is not always possible, since not all optically isomeric compounds give two types of crystal.

(2) A suitable mould, when grown in a solution containing the racemic mixture, may consume one enantiomer more rapidly than the other; this other enantiomer is therefore left behind.
(3) If a racemic acid is reacted with one enantiomer, e.g., the (+) form of an optically active base, two salts are formed, i.e., one from the (+) acid and the (+) base, and one from the (−) acid and the (+) base. These are no longer enantiomers, but are called diastereoisomers, and have different physical properties, e.g., different solubilities, etc. Hence the salts can be separated by fractional crystallization, and the free acids then liberated by treating with strong acids.

It should be noted that when a compound, which is capable of showing optical isomerism, is synthesized in the laboratory, the racemic mixture is obtained. This can be seen in the synthesis of the cyanohydrin (12) and (14) from ethanal (13) and hydrogen cyanide. It can be seen that attack by the cyanide

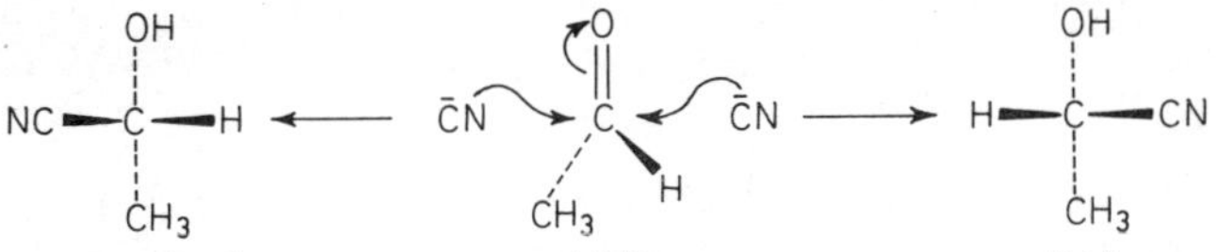

(12) (13) (14)

group can occur equally well from either side (and statistically, half will attack each side). In contrast to this, compounds formed in nature which are capable of optical activity are found in an optically active form, since they are produced *via* a specific route.

14.5 Homologous Series and Physical Properties

A set of compounds having the same functional groups and differing from each other only by an integral number of $-CH_2-$ units is called a HOMOLOGOUS SERIES. For example, the alkanes, general formula C_nH_{2n+2}, constitute a homologous series. Trends can be seen within the series, such as the increasing melting and boiling points as the series is ascended owing to the increasing relative molecular masses and van der Waals forces (p. 24).

Another factor which can affect the melting and boiling points of compounds is shape. For example, consider pentane (15) and its isomer 2,2-dimethylpropane (16). Although they have the same relative molecular mass (since they

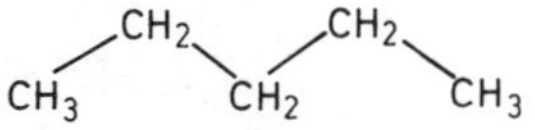

(15)

Pentane
Relative molecular mass 72
Boiling point 36°C

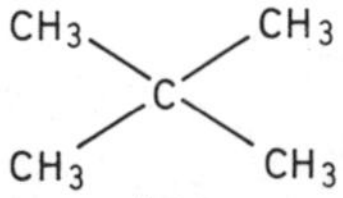

(16)

2,2-Dimethylpropane
Relative molecular mass 72
Boiling point 9°C

are isomers), pentane has the higher boiling point. This is because the latter has a more extended shape and a large surface area; hence there is appreciable interaction between chains or molecules, giving it the higher boiling point.

The presence of polar groups on adjacent molecules also has the effect of increasing intermolecular forces (*see* Chapter 3) and hence boiling points, etc. With the alcohols, for example, hydrogen bonding can also occur because of the hydroxyl groups. Hydrogen bonding can also affect viscosity and it can be shown that, generally speaking, as the number of hydroxyl groups in a compound increases, so does the viscosity. Another example of hydrogen bonding includes the dimerization of acids (*see* p. 26) in non-polar solvents.

Finally, interaction between atoms or groups within the same molecule also has an effect on properties, such as shape. For example, cyclohexane has the structural formula (17). The structure is not planar, but has been shown to exist

$$\begin{array}{ccc} & CH_2 & \\ CH_2 & & CH_2 \\ | & & | \\ CH_2 & & CH_2 \\ & CH_2 & \end{array}$$

(17)

in 'chair' (18) and 'boat' (19) forms, which allow the bond angles to approach the tetrahedral angle more closely. The two forms are so readily interconvertible

(18) Chair form

(19) Boat form

that neither one can be isolated; however the chair form predominates since it allows the hydrogen atoms to be further apart. (*See* also intramolecular forces in proteins: Chapter 16.)

14.6 Some Basic Principles

The cleavage of covalent bonds is the fundamental process of organic reactions, and can take place in one of the following ways:

$$R{:}X \rightarrow R\cdot + \cdot X \quad \text{(homolytic fission)}$$

$$R{:}X \rightarrow R{:}^- + X^+ \quad \text{or} \quad R^+ + {:}X^- \quad \text{(heterolytic fission)}$$

In the first situation, each atom separates with one electron, to give highly reactive species called free radicals (*see* also Chapter 10, p. 85) – their high reactivity is because of their unpaired electron. This process is termed HOMOLYTIC FISSION, and such reactions tend to occur in the gas phase and in non-polar solvents. Alternatively, one atom may hold on to both of the electrons (or neither of them), hence producing positive and negative ions – these reactions tend to occur more readily in polar solvents. If the positive charge is carried on the carbon, it is referred to as a CARBONIUM ION; if it is negative, it is called a CARBANION.

It should be noted that during homolytic fission, the movement of a single electron can be depicted by a half arrow, i.e.,

$$R{-}X \longrightarrow R\bullet + X\bullet$$

14.6.1 ELECTROPHILES AND NUCLEOPHILES

In the Introduction (p. 12) it was mentioned how covalent bonds can become polarized because of the different electronegativities of the atoms joined by the bond. Hence, in a carbon–chlorine bond, for example (*Figure 12*), the carbon atom becomes slightly electron deficient ($\delta+$) and hence will attract NUCLEOPHILES, i.e., species which have one or more unshared pairs of electrons (e.g., OH^-, CN^-, $:NH_3$, etc.) which they can donate to form a new bond with the atom which they are attacking. Hence the carbon atom is said to undergo NUCLEOPHILIC ATTACK, and if we let the nucleophile be denoted as B^-, we can write a MECHANISM for this attack. The curved arrows indicate the movement of electron pairs; the tail of the arrow indicates the origin of the electron pair, and the head its destination:

$$\gt\overset{\delta+}{C}{-}\overset{\delta-}{Cl} + B^- \longrightarrow \gt C{-}B + Cl^-$$

The electron-deficient sites (in the above case, the $\delta+$ carbon) or positively charged entities (*see* Chapter 15) which can be attacked by nucleophiles are known as ELECTROPHILES. Reactions can be classified as nucleophilic substitutions, nucleophilic additions, electrophilic additions, and so on. This will be illustrated by the examples given in Chapter 15.

The inductive effect (*see* p. 12) in the above example is a 'pull' of electrons towards the chlorine and away from the carbon, and this is described as a negative inductive effect (or $-I$ effect). However, alkyl groups (*see* p. 152) have an electron 'pushing' or $+I$ effect, and can be represented as shown in structure (20).

$$CH_3 \rightarrow \underset{\underset{CH_3}{\uparrow}}{\overset{\overset{CH_3}{\downarrow}}{C}} \rightarrow Cl$$

(20)

This $+I$ effect of alkyl groups has an important effect on the stability of carbonium ions (*see* Chapter 15).

15

A STUDY OF SOME FUNCTIONAL GROUPS

15.1 Alkanes

15.1.1 GENERAL PROPERTIES

The alkanes constitute a homologous series having the general formula C_nH_{2n+2}. The first ten are named in *Table 15.1.* Note that the first four have trivial names whereas the rest have the number of carbon atoms indicated by the name. If a

Table 15.1 The alkanes

CH_4	methane	C_6H_{14}	hexane
C_2H_6	ethane	C_7H_{16}	heptane
C_3H_8	propane	C_8H_{18}	octane
C_4H_{10}	butane	C_9H_{20}	nonane
C_5H_{12}	pentane	$C_{10}H_{22}$	decane

hydrogen atom is removed from an alkane an ALKYL GROUP is formed, and is named by removing '-ane' from the name of the alkane and replacing it with '-yl', e.g.,

CH_3-	C_2H_5-	C_3H_7-
methyl	ethyl	propyl

Alkyl groups are often given the symbol R for brevity or for generalizing.

In naming branched or substituted alkanes, the following general rules are used:

(1) The longest straight chain is found and is numbered consecutively from the end which gives the substituents the lowest possible number.

(2) Each substituent is specified by name and number (*see* below); if more than one substituent is present, they are written in alphabetical order, as in examples (1)–(3).

$$\overset{1}{CH_3}-\underset{\displaystyle CH_3}{\overset{2}{CH}}-\overset{3}{CH_2}-\overset{4}{CH_3}$$

(1) 2-Methylbutane (NOT 3-methylbutane)

$$\overset{1}{CH_3}-\underset{\displaystyle CH_3}{\overset{\displaystyle CH_3}{\overset{2}{C}}}-\overset{3}{CH_2}-\overset{4}{CH_2}-\overset{5}{CH_3}$$

(2) 2,2-Dimethylpentane

$$\overset{1}{CH_3}-\underset{\displaystyle CH_3}{\overset{2}{CH}}-\underset{\displaystyle C_2H_5}{\overset{3}{CH}}-\overset{4}{CH_2}-\overset{5}{CH_2}-\overset{6}{CH_2}-\overset{7}{CH_2}-\overset{8}{CH_3}$$

(3) 3-Ethyl-2-methyloctane (NOT 2-methyl-3-ethyloctane)

The alkanes occur naturally as natural gas (mainly methane together with smaller amounts of the other gaseous alkanes, i.e., ethane, propane and butane) and as crude petroleum. Crude petroleum (formed from the remains of marine and plant life over many millions of years) is made up of many alkanes up to about C_{40}, together with cycloalkanes, aromatic hydrocarbons, and impurities. On fractionally distilling crude petroleum, it is split up into various fractions. Fractions C_5–C_7 are used as solvents and light petroleum, C_6–C_{11} as motor fuel, C_{12}–C_{16} as paraffin (kerosene), C_{13}–C_{18} as fuel oil, and greater than C_{18} for lubricating oils and greases, waxes, etc.

The liquid alkanes are oily, insoluble in water but soluble in organic solvents. They can be prepared in the laboratory by heating the sodium salt of a carboxylic acid with soda lime (NaOH–CaO), e.g.,

$$CH_3COONa + NaOH \rightarrow CH_4 + Na_2CO_3$$

The alkanes are fairly unreactive since they are saturated, i.e., contain no double or triple bonds but only strong C–H and C–C bonds.

15.1.2 REACTIONS OF THE ALKANES

15.1.2.1 Oxidation

Alkanes burn readily in air to form carbon dioxide and water, e.g.,

$$CH_4 + 2O_2 \rightarrow CO_2 + 2H_2O$$

Many of the alkanes are therefore used as fuels of one form or another (*see* above).

15.1.2.2 Chlorination

In direct sunlight, methane and chlorine can explode, depositing carbon:

$$CH_4 + 2Cl_2 \rightarrow C + 4HCl$$

In diffused light, chlorinated products are formed by a free radical process (*see* Chapter 10, p. 85):

$CH_4 + Cl_2 \rightarrow HCl + CH_3Cl$ chloromethane (methyl chloride)
$CH_3Cl + Cl_2 \rightarrow HCl + CH_2Cl_2$ dichloromethane (methylene chloride)
$CH_2Cl_2 + Cl_2 \rightarrow HCl + CHCl_3$ trichloromethane (chloroform)
$CHCl_3 + Cl_2 \rightarrow HCl + CCl_4$ tetrachloromethane (carbon tetrachloride)

15.1.2.3 Sulphonation

Treating the higher alkanes with fuming sulphuric acid gives sulphonic acids:

$$RH + H_2SO_4 \rightarrow \underset{\text{Alkanesulphonic acid}}{R.SO_3H} + H_2O$$

Since the acidity of an acid, HA, is determined by both the electronegativity of A and the stability of the anion, A^-, the alkanesulphonic acids are strong monobasic acids since (1) the presence of a large 'electron withdrawing' group weakens the O–H bond and (2) the resultant anion is stabilized by resonance:

$$R{-}S(=O)_2{-}O^- \longleftrightarrow R{-}S(=O)(O^-){=}O \longleftrightarrow R{-}S(=O)_2{-}O^-$$

(4) (5) (6)

The sodium salts of these acids are used as detergents, since the alkyl chain is oil-soluble and the sulphonate group is water-soluble; hence the detergent, in aqueous solution, can remove greasy stains, etc.

15.1.2.4 Nitration

Alkanes in the gas phase when treated with nitric acid vapour form a mixture of nitroalkanes, since various hydrogen atoms are replaced by the nitro-group, $-NO_2$ (and also some degradation of the chain occurs):

$$RH + HNO_3 \rightarrow \underset{\text{Nitroalkane}}{R.NO_2} + H_2O$$

The nitroalkanes are used as solvents and in synthetic work.

15.1.2.5 Cracking

Lighter fractions (for petrol) and short chain alkenes (e.g., ethene and propene for use in the plastics industry – *see* Chapter 17) can be obtained by 'cracking' or breaking down the larger molecules in the heavier fractions (by homolytic fission); catalysts include oxides of silicon, aluminium and thorium.

$$C_nH_{2n+2} \rightarrow C_yH_{2y+2} + C_zH_{2z} \quad (\text{where } y + z = n)$$

15.2 Alkenes

15.2.1 GENERAL PROPERTIES

The alkenes constitute a homologous series of general formula C_nH_{2n}. The appropriate suffix is now '-ene'. In naming the alkenes, the chain is numbered so that the double bond has the lowest number possible, as shown in examples (7)–(11).

Alkenes are obtained on a large scale by the cracking of petroleum (*see* before). In the laboratory, they can be prepared by the dehydration of alcohols (*see* p. 167) or the dehydrohalogenation of halogenoalkanes (*see* p. 165).

The most important feature of alkenes is the double bond (*see* Chapter 2, p. 21, for structure). Hence, most of the reactions undergone by alkenes are due

$CH_2{=}CH_2$
Ethene (ethylene)
(7)

$CH_3.CH{=}CH_2$
Propene (propylene)
(10)

$\overset{4}{C}H_3.\overset{3}{C}H_2.\overset{2}{C}H{=}\overset{1}{C}H_2$
But-1-ene
(8)

$CH_3.CH{=}CH.CH_3$
But-2-ene
(11)

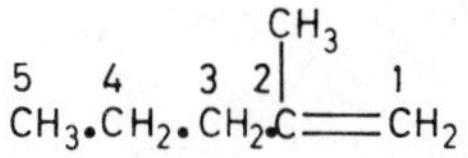

$\overset{5}{C}H_3.\overset{4}{C}H_2.\overset{3}{C}H_2.\overset{2}{C}(CH_3){=}\overset{1}{C}H_2$ 2-Methylpent-1-ene
(9)

mainly to the double bond, and are generally electrophilic additions, i.e., additions by electron deficient species. The alkenes, therefore, are unsaturated and reactive (unlike the alkanes).

15.2.2 REACTIONS OF THE ALKENES

15.2.2.1 Hydrogenation

Alkenes combine with hydrogen in the presence of a catalyst to form alkanes:

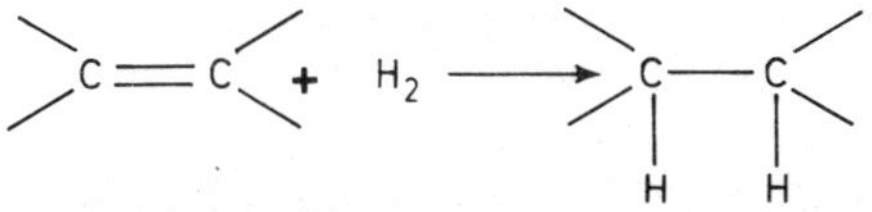

$$>C{=}C< + H_2 \longrightarrow >\underset{H}{C}-\underset{H}{C}<$$

Catalysts which can be used include finely divided platinum, palladium, and Raney nickel (prepared by removing aluminium from a Ni/Al alloy with caustic soda) at room temperature, or nickel at 200–300 °C. This process is used in hydrogenating double bonds present in vegetable oils to give solid fats during the manufacture of margarine.

15.2.2.2 Halogenation

Chlorine and bromine both readily add to alkenes (the decolourization of a solution of bromine in tetrachloromethane in the cold is a test for a double bond). This reaction is electrophilic addition. When ethene, for example, is approached by a bromine molecule, the high electron density of the double bond induces polarization of the bromine molecule, hence creating an electrophilic site (a δ+ bromine):

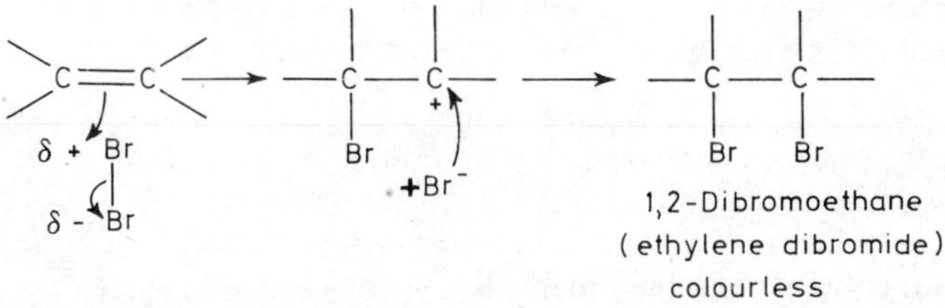

$$>C{=}C< \;\; \overset{\delta+}{Br}-\overset{\delta-}{Br} \longrightarrow -\underset{Br}{\overset{|}{C}}-\overset{+}{\overset{|}{C}}- \;\; +Br^- \longrightarrow -\underset{Br}{\overset{|}{C}}-\underset{Br}{\overset{|}{C}}-$$

1,2-Dibromoethane
(ethylene dibromide)
colourless

With chlorine or bromine water, HOX adds on, e.g.,

$$>C=C< \longrightarrow -\overset{|}{\underset{Br}{C}}-\overset{|}{\overset{+}{C}}- + HO^- \longrightarrow -\overset{|}{\underset{Br}{C}}-\overset{|}{\underset{OH}{C}}-$$

$\delta+$ Br — $\delta-$ OH

2-Bromoethanol
(ethylene bromohydrin)

15.2.2.3 Addition of Hydrogen Halides

Hydrogen halides, HX, add readily to alkenes; hydrogen iodide is the most reactive whilst hydrogen bromide and chloride are less so, e.g.,

$$>C=C< \longrightarrow -\overset{|}{\underset{H}{C}}-\overset{|}{\overset{+}{C}}- + Br^- \longrightarrow -\overset{|}{\underset{H}{C}}-\overset{|}{\underset{Br}{C}}-$$

$\delta+$ H — $\delta-$ Br

Bromoethane
(ethyl bromide)

In the case of an unsymmetrical alkene, such as propene, two products could be formed, i.e., $CH_3.CH_2.CH_2Br$ or $CH_3.CHBr.CH_3$. The latter product (2-bromopropane) is in fact formed, as predicted by MARKOWNIKOFF'S RULE, which states that: In the addition of an unsymmetrical adduct to an unsymmetrical alkene, the more negative part of the adduct adds on to the unsaturated carbon atom having the least number of hydrogen atoms. The two possible routes are now shown:

$$CH_3{\cdot}CH{=}CH_2$$

$\delta+$ H — $\delta-$ Br

$$CH_3{\cdot}CH_2{\cdot}\overset{+}{C}H_2 \longleftarrow \quad \longrightarrow CH_3{\cdot}\overset{+}{C}H{\cdot}CH_3$$

(Primary carbonium ion) (Secondary carbonium ion)

The secondary carbonium ion is, in fact, the more stable since it has two alkyl groups 'pushing' electrons (p. 151) towards the positively charged carbon, hence tending to reduce its charge density, i.e., the more alkyl groups there are on the positive carbon of a carbonium ion, the more stable and the more easily formed that carbonium ion will be. In the case of the primary carbonium ion, there is only one alkyl group (the ethyl group). Hence, the final product is 2-bromopropane.

Note that if traces of peroxides are present, then the above HBr reaction can follow a free radical mechanism and anti-Markownikoff addition occurs.

15.2.2.4 Polymerization

The polymerization of alkenes and their derivatives is discussed in Chapter 17.

15.2.2.5 Oxidation

Alkenes burn in air to give carbon dioxide and water, e.g.,

$$C_2H_4 + 3O_2 \rightarrow 2CO_2 + 2H_2O$$

Alkenes are oxidized by alkaline potassium manganate(VII) (permanganate) solution, in the cold, to give a dihydroxy compound. The manganate(VII) first turns green (MnO_4^{2-}) and then brown (MnO_2), and this can be used as a test for unsaturation:

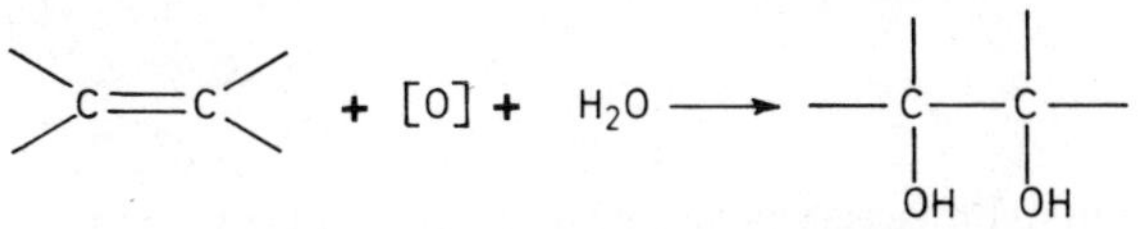

$$>C{=}C< \; + [O] + H_2O \longrightarrow -\overset{|}{\underset{OH}{C}}-\overset{|}{\underset{OH}{C}}-$$

Ethane-1,2-diol
(ethylene glycol)

With peroxocarboxylic (percarboxylic) acids, alkenes are oxidized to alkene oxides or epoxides, e.g.,

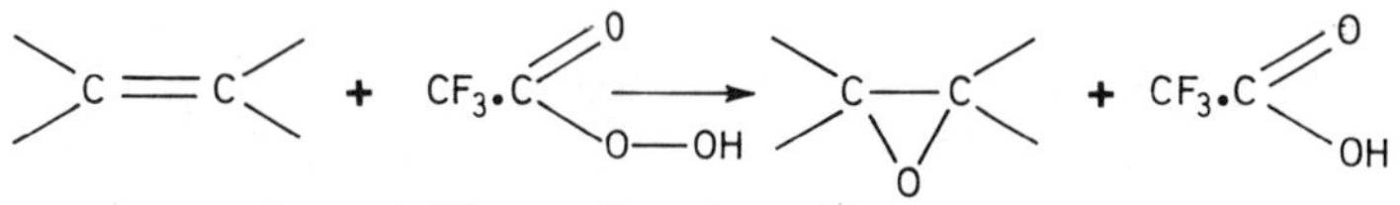

$$>C{=}C< \; + CF_3.C(=O)O{-}OH \longrightarrow >C{-}C< \text{(epoxide, bridged by O)} + CF_3.C(=O)OH$$

Peroxotrifluoroethanoic acid

Alkenes are oxidized by trioxygen (ozone) to form ozonides. This is done by passing the trioxygen into a solution of the alkene in an inert solvent (e.g., tetrachloromethane); the ozonide formed is either hydrolysed with water or reduced to produce carbonyl compounds (see p. 170):

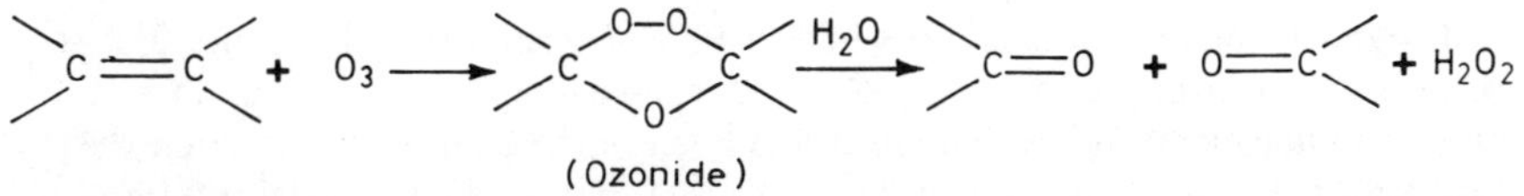

$$>C{=}C< \; + O_3 \longrightarrow >C(O{-}O)(O)C< \xrightarrow{H_2O} >C{=}O + O{=}C< + H_2O_2$$

(Ozonide)

By identifying the two carbonyl compounds, valuable structural information can be obtained concerning the original alkene; hence ozonolysis is used in structure determination of alkenes.

15.2.2.6 Hydration

The addition of the elements of water across a double or triple bond is called a HYDRATION reaction. This addition reaction is performed in aqueous sulphuric acid (60–80%) followed by further dilution of the mixture with water; the net result is an alcohol, and this process is used in industry for producing the latter.

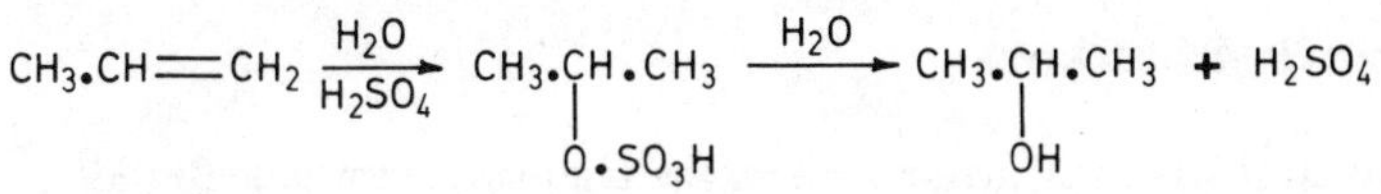

$$CH_3.CH{=}CH_2 \xrightarrow[H_2SO_4]{H_2O} CH_3.\underset{O.SO_3H}{\underset{|}{C}H}.CH_3 \xrightarrow{H_2O} CH_3.\underset{OH}{\underset{|}{C}H}.CH_3 + H_2SO_4$$

15.2.3 INDUSTRIAL USES OF ALKENES

Alkenes are of great use in industry. For example, ethene is used: in the production of poly(ethene) (polythene); in the production of epoxyethane (ethylene oxide) followed by acid hydrolysis to form ethane-1,2-diol (ethylene glycol), used in the antifreeze industry and for making terylene; in the production of ethanol (*see* above); in the production of chloroethene (vinyl chloride) and other substituted ethene derivatives (*see* Chapter 17) for use in the plastics industry; for ripening fruit.

15.3 Alkynes

The alkynes constitute a homologous series having the general formula C_nH_{2n-2}. The rules for nomenclature are the same as for alkenes, except that the suffix is now '-yne', e.g., structures (12) and (13). The characteristic of the alkynes is the

$$CH{\equiv}CH \qquad\qquad CH_3.C{\equiv}CH$$

Ethyne (acetylene) (12) Propyne (methylacetylene) (13)

triple bond (*see* Chapter 2, p. 22, for structure). Ethyne can be prepared in the laboratory by adding water to calcium dicarbide (carbide):

$$CaC_2 + 2H_2O \rightarrow Ca(OH)_2 + C_2H_2$$

Alkynes, like alkenes, are unsaturated and undergo many addition reactions readily; these are very similar to those of the alkenes, but occur twice, e.g.,

$$CH{\equiv}CH + Cl_2 \longrightarrow \underset{\text{1, 2-Dichloroethene}}{CHCl{=}CHCl} \xrightarrow{Cl_2} \underset{\text{1, 1, 2, 2-Tetrachloroethane}}{CHCl_2.CHCl_2}$$

Ethyne differs from the alkenes in that its hydrogen is slightly acidic. This is because the electron pair in the C–H bond is, on average, further away from the other bonding electrons (which are constrained in the triple bond between the two carbon nuclei) than in alkenes (or alkanes). Hence, the C–H electron pair in ethyne can move closer to the carbon nucleus than is normal for a hydrocarbon and hence the hydrogen is more easily removed by proton acceptors. Therefore, if ethyne is passed into ammoniacal silver nitrate solution, a white precipitate of silver dicarbide (acetylide) is formed:

$$C_2H_2 \xrightarrow[\text{silver nitrate}]{\text{Ammoniacal}} Ag_2C_2\downarrow$$

This serves to distinguish ethyne from alkenes, since the latter do not give this reaction.

Ethyne is used for welding (oxyacetylene welding) and as a starting material for the manufacture of plastics and grease solvents, etc.

15.4 Aromatic Hydrocarbons

In Chapter 2 (p. 22) it was explained how benzene is a planar molecule (bond angles 120 degrees) and how delocalization of the 2*p* electrons leads to stability (*see* also Chapter 5, p. 43, for experimental evidence). Because of this stability, most of benzene's reactions are those of substitution; in this way, the aromatic nature of the ring is preserved (whereas addition reactions would destroy this) – see below.

For convenience, benzene is often written as a single Kekulé form [*Figure 2.5*(*b*)] or as its delocalized form [*Figure 2.5*(*c*)], i.e., (14a) and (14b). Nomenclature of substituted benzene compounds is based on a numbering system where

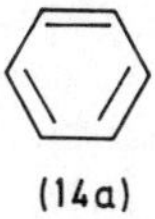

(14a) or (14b)

the carbon atoms of the ring are numbered so as to give the substituents the lowest possible number, as shown in *Table 15.2.*

Table 15.2 Substituents in the benzene ring

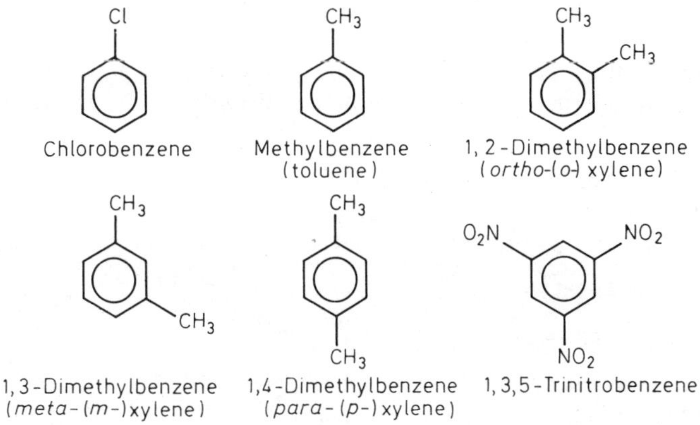

Cl — Chlorobenzene

CH_3 — Methylbenzene (toluene)

CH_3, CH_3 — 1, 2-Dimethylbenzene (*ortho*-(*o*-) xylene)

CH_3, CH_3 — 1, 3-Dimethylbenzene (*meta*-(*m*-)xylene)

CH_3, CH_3 — 1,4-Dimethylbenzene (*para*-(*p*-)xylene)

O_2N, NO_2, NO_2 — 1, 3,5-Trinitrobenzene

15.4.1 ADDITION REACTIONS

Benzene does give a few addition reactions. With hydrogen (in the presence of a nickel or platinum black catalyst at about 200 °C), it forms cyclohexane, C_6H_{12}:

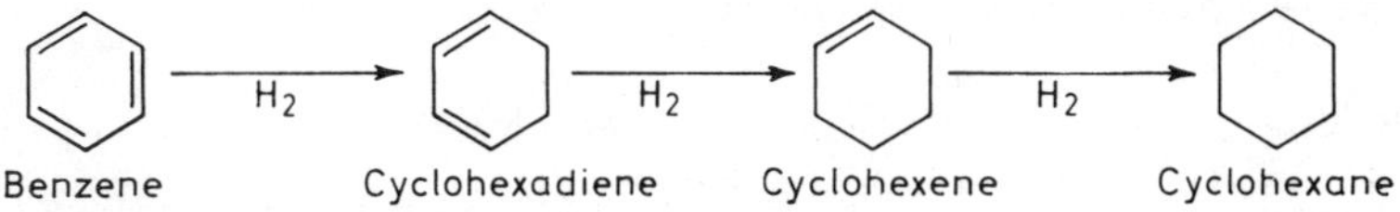

Benzene $\xrightarrow{H_2}$ Cyclohexadiene $\xrightarrow{H_2}$ Cyclohexene $\xrightarrow{H_2}$ Cyclohexane

With chlorine under irradiation with sunlight or u.v. light, addition occurs to give eventually 1,2,3,4,5,6-hexachlorocyclohexane (benzene hexachloride):

$$C_6H_6 + 3Cl_2 \rightarrow C_6H_6Cl_6$$

15.4.2 SUBSTITUTION REACTIONS

The majority of the reactions of benzene are electrophilic substitutions, the general mechanism being:

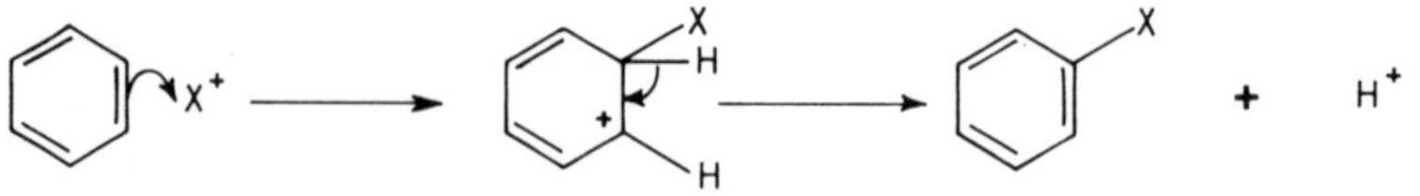

Note that the positive charge produced in the ring can be delocalized as in structures (15), (16) and (17), so it is often written as (18).

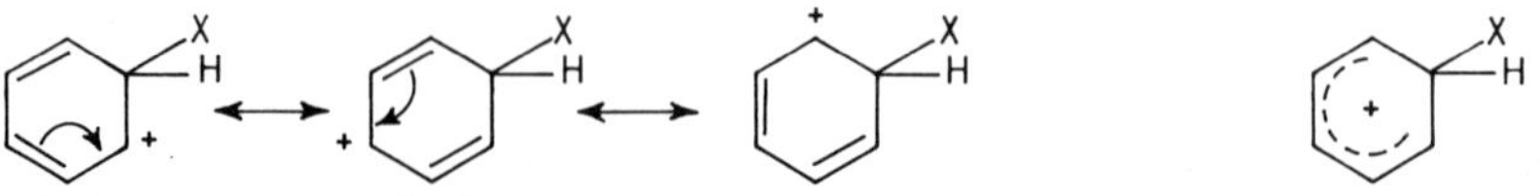

(15) (16) (17) (18)

15.4.2.1 Nitration

On warming benzene with a mixture of concentrated nitric and sulphuric acids, nitrobenzene is produced. The sulphuric acid generates the nitryl cation (nitronium ion), NO_2^+.

$$2H_2SO_4 + HNO_3 \rightarrow NO_2^+ + H_3O^+ + 2HSO_4^-$$

Then:

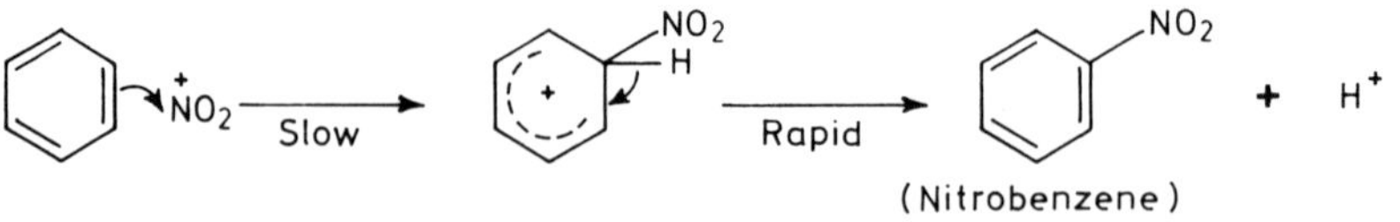

Slow Rapid

(Nitrobenzene)

This reaction occurs at temperatures up to 60 °C; if the temperature is allowed to rise to 100 °C, a second nitro-group is introduced to form 1,3-dinitrobenzene (*m*-dinitrobenzene).

With methylbenzene (toluene), nitration gives a mixture of methyl-2-nitrobenzene (19) and methyl-4-nitrobenzene (20) (*o*- and *p*-nitrotoluene).

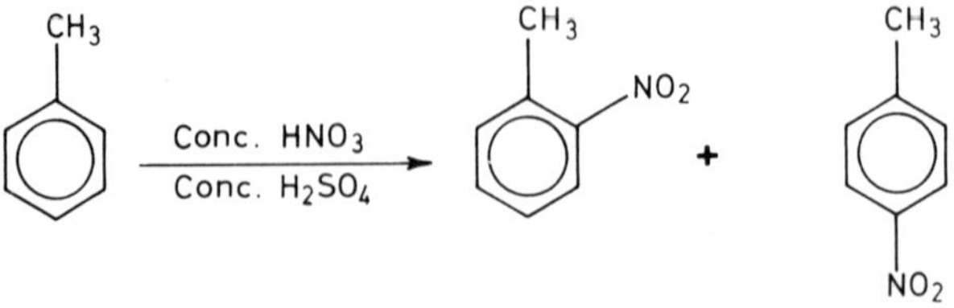

Conc. HNO_3 / Conc. H_2SO_4

(19) (20)

The positions at which additional substituents enter the ring depend on which substituents are already present. It is observed that electron withdrawing groups attached to the ring (e.g., $-NO_2$) give rise to 3- (*m*-) substitution, whilst electron pushing or donating groups attached to the ring (e.g., $-CH_3$) give rise to 2- and

4- (*o*- and *p*-) substitutions. Note that $-OH$, $-NH_2$, and halogens, when attached to a benzene ring, are 2- and 4-directing, since non-bonding pairs of electrons in their *p* orbitals can be delocalized in the ring, e.g., structures (21)–(23) and so on.

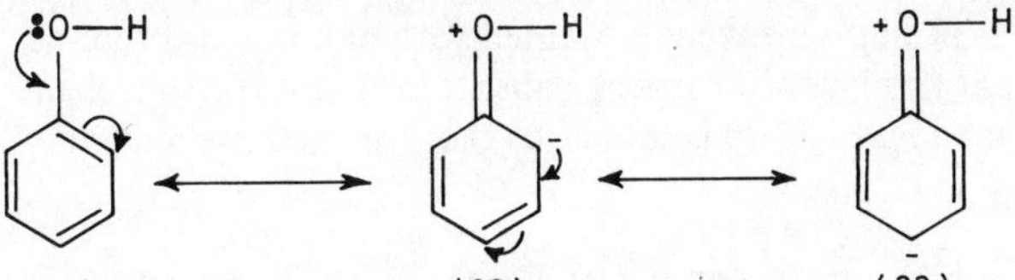

(21) (22) (23)

It can be seen that the 2- and 4-positions becomes richer in electron density, and so are more reactive towards electrophiles.

15.4.2.2 Sulphonation

With fuming sulphuric acid [sulphuric acid containing sulphur(VI) oxide, SO_3] at room temperature, or concentrated sulphuric acid at elevated temperature, benzene is sulphonated to give benzenesulphonic acid (24). The reaction is reversible.

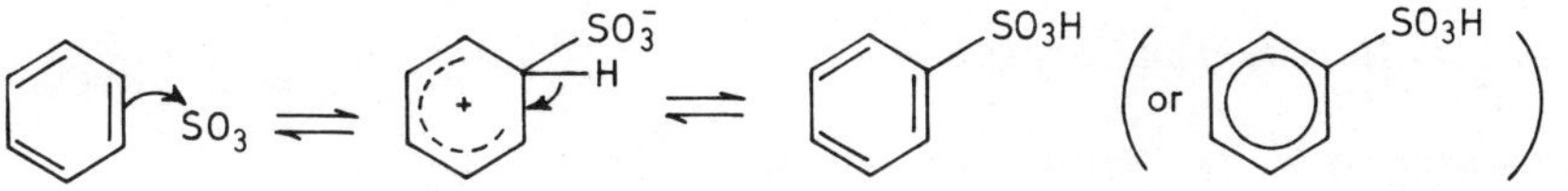

(24)

With methylbenzene, a mixture of 2- and 4-methylbenzenesulphonic acids (*o*- and *p*-toluenesulphonic acids) (25) and (26) are formed.

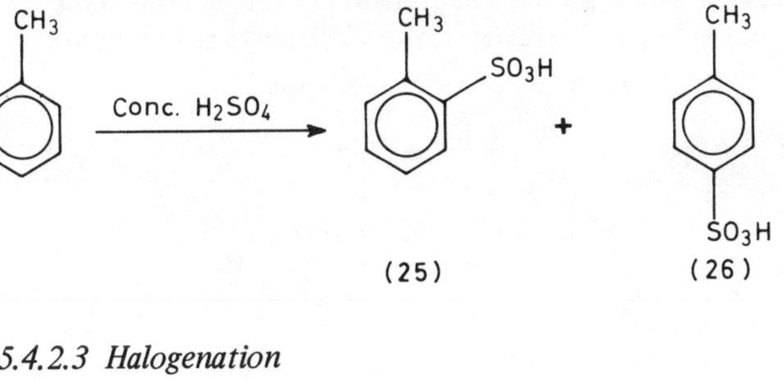

(25) (26)

15.4.2.3 Halogenation

Substitution into the ring occurs when benzene or methylbenzene is reacted with chlorine or bromine in the presence of catalysts called halogen carriers, e.g., $AlCl_3$, $FeCl_3$, or iodine. The function of the halogen carrier is to create a stronger electrophile. In the case of the aluminium halide, the electron deficient aluminium (*see* Chapter 12, p. 98) acts as a Lewis acid, i.e., it accepts an electron pair; the iron(III) halides can also do this because of vacant *d* orbitals. The mechanism, therefore, is as follows:

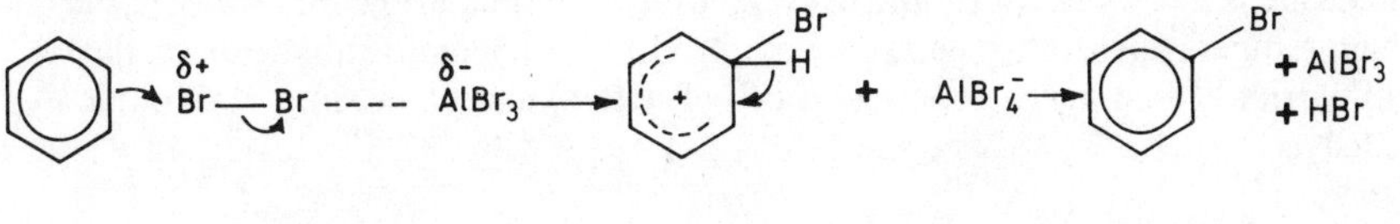

In the case of iodine, it can act as a halogen carrier because its outer electrons are far from the nucleus and are therefore easily displaced, e.g., iodine is easily polarized and this, in turn, can polarize the chlorine or bromine, hence creating an electrophilic site.

It should be noted that if methylbenzene is treated with chlorine or bromine in the presence of u.v. light (no halogen carrier), substitution into the side chain occurs; the mechanism (*see Figure 15.1*) is similar to that for methane and chlorine, i.e., free radical (p. 85).

$$C_6H_5.CH_3 + Cl_2 \longrightarrow C_6H_5.CH_2Cl + HCl \xrightarrow{Cl_2} C_6H_5.CHCl_2 + HCl$$

(Chloromethyl)benzene (benzyl chloride) — (Dichloromethyl)benzene (benzal chloride)

$$\xrightarrow{Cl_2} C_6H_5.CCl_3 + HCl$$

(Trichloromethyl)benzene (benzotrichloride)

Figure 15.1

15.4.2.4 Alkylation and Acylation

Arenes (aromatic hydrocarbons) are alkylated when treated with halogenoalkanes (alkyl halides) (*see* p. 163) in the presence of Lewis acids, e.g., $AlCl_3$:

$$C_6H_6 + RCl \xrightarrow{AlCl_3} C_6H_5.R + HCl$$

This is called the Friedel–Crafts reaction. The function of the catalyst is the same as in the halogenation reaction. However, it is difficult to stop more than

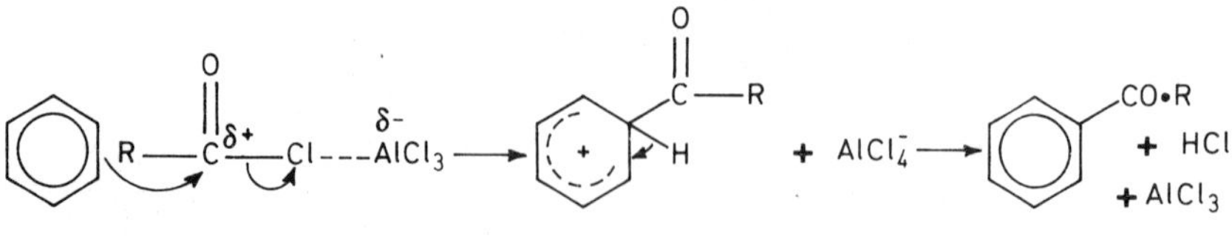

$$C_6H_6 + R{-}\overset{\overset{O}{\|}}{C}{}^{\delta+}{-}Cl{\cdots}\overset{\delta-}{AlCl_3} \longrightarrow [C_6H_5(H)(C(=O){-}R)]^+ + AlCl_4^- \longrightarrow C_6H_5.CO.R + HCl + AlCl_3$$

Figure 15.2

one alkyl group from entering the benzene ring; if an acyl chloride (*see* p. 178) is used instead of a halogenoalkane, this problem does not arise. This acylation reaction produces a ketone, e.g., *Figure 15.2.*

15.4.3 OXIDATIONS

Benzene is not oxidized by the usual laboratory oxidizing agents. However, at high temperatures, benzene can be oxidized to butenedioic anhydride (maleic anhydride) (used in the production of polyesters) using a vanadium(V) oxide catalyst:

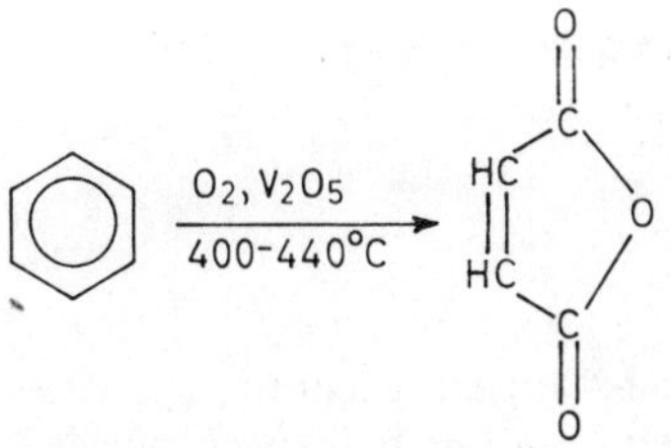

$$C_6H_6 \xrightarrow[400-440°C]{O_2, V_2O_5} \text{maleic anhydride}$$

The side chains of arenes, however, are readily oxidized to an acid group attached to the ring (no matter how long the side chain is) by refluxing with acid or alkaline potassium manganate(VII) solution, acidified potassium dichromate(VI) solution, etc.:

$$C_6H_5.CH_3 + 3[O] \rightarrow C_6H_5.COOH + H_2O$$

Benzenecarboxylic acid (benzoic acid)

15.5 The Carbon–Halogen Bond

Halogenoalkanes (alkyl halides) constitute a homologous series of general formula $C_nH_{2n+1}X$ where X = Cl, Br or I. They are systematically named as halogen substituted alkanes. Halogenobenzenes are named as halogen substituted benzenes (*see Table 15.3*).

Table 15.3 Alkyl and aryl halides

CH_3Cl Chloromethane (methyl chloride)		$CH_3.CH_2Br$ Bromoethane (ethyl bromide)
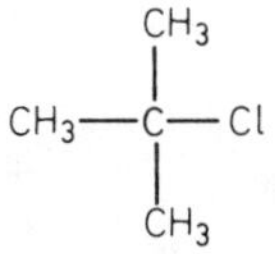 $(CH_3)_3C-Cl$ 2-Chloro-2-methylpropane (t-butyl chloride)	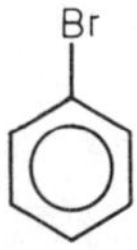 C_6H_5Br Bromobenzene	 C_6H_5Cl Chlorobenzene

Halogenoalkanes can be prepared as follows:

(1) By the reaction of alcohols with phosphorus halides, e.g.,

$$3C_2H_5OH + PCl_3 \rightarrow 3C_2H_5Cl + H_3PO_3$$

$$C_2H_5OH + PCl_5 \rightarrow C_2H_5Cl + HCl + POCl_3$$

(2) By the action of hydrogen halides on alcohols (see p. 168).

Halogenobenzenes can be prepared by the reaction of a halogen with benzene using a halogen carrier (*see* p. 161).

Because of the inductive effect (e.g., $C_2H_5 \rightarrow Cl$), halogenoalkanes are polarized, and therefore the resultant $\delta+$ carbon makes them susceptible to nucleophilic attack (p. 151). In general

$$\overset{\delta+}{>C}-\overset{\delta-}{X} + B^- \longrightarrow >C-B + X^-$$

This type of reaction is known as a nucleophilic substitution reaction; the nucleophile, B^-, could be OH^-, for example, from aqueous sodium hydroxide, and X^- would be a halide ion. Halogenobenzenes, however, are more reluctant to undergo nucleophilic substitution reactions. This is because the high electron density of the benzene ring (or nucleus) inhibits the approach of nucleophiles (there is also the possibility of *p* orbital overlap between the halogen and the ring, which would tend to strengthen the bond). Therefore, chlorobenzene, for example, will undergo substitution by OH^- only when the reaction conditions are vigorous:

$$C_6H_5Cl \xrightarrow[300°C/\text{pressure}]{\text{Aqueous NaOH}} C_6H_5OH$$

It should be noted that the order of reactivity of the halogenoalkanes towards substitution by nucleophiles is

$$RI > RBr > RCl$$

where R is the alkyl group.

Finally, it should be mentioned that the mechanisms of hydrolysis of primary and tertiary halogenoalkanes are very different, and these are discussed in Chapter 10 (p. 85). This difference reflects the influence of the three (electron pushing) alkyl groups in the case of the tertiary compound.

15.5.1 REACTIONS OF HALOGENOALKANES

(1) Halogenoalkanes are hydrolysed to alcohols when refluxed with aqueous alkali, e.g.,

$$C_2H_5Br + NaOH_{(aq)} \rightarrow \underset{\substack{\text{Ethanol}\\ \text{(ethyl alcohol)}}}{C_2H_5OH} + NaBr$$

or

$$H-\underset{H}{\overset{H}{C}}-\underset{H}{\overset{H}{\overset{\delta+}{C}}}-\overset{\delta-}{Br} \quad {}^-OH \longrightarrow H-\underset{H}{\overset{H}{C}}-\underset{H}{\overset{H}{C}}-O-H + Br^-$$

However, in alcoholic solution, halogenoalkanes and sodium or potassium hydroxide give elimination reactions to form alkenes, e.g.,

$$CH_3-\underset{H}{\overset{Br}{C}}-\underset{H}{\overset{H}{C}}-H \quad {}^-OH \longrightarrow CH_3.CH=CH_2 + H_2O + Br^-$$

(2) Refluxing a halogenoalkane in aqueous alcohol with sodium or potassium cyanide gives a nitrile (cyanide), e.g.:

$$C_2H_5Br + KCN \rightarrow \underset{\substack{\text{Propanonitrile}\\ \text{(ethyl cyanide)}}}{C_2H_5CN} + KBr$$

(3) Refluxing the sodium salt of a carboxylic acid with a halogenoalkane gives an ester, e.g.:

$$C_2H_5Br + CH_3.COONa \rightarrow \underset{\substack{\text{Ethyl ethanoate}\\ \text{(ethyl acetate)}}}{CH_3.COOC_2H_5} + NaBr$$

(4) On heating an alcoholic solution of a halogenoalkane with ammonia in a sealed vessel, an amine salt is produced (*see also* p. 183).

$$C_2H_5Br + NH_3 \rightarrow \underset{\substack{\text{Ethylammonium}\\ \text{bromide}\\ \text{(ethylamine hydrobromide)}}}{C_2H_5\overset{+}{N}H_3\overset{-}{Br}}$$

(5) Refluxing a halogenoalkane with an alcoholic solution of an alkoxide gives an ether (Williamson's synthesis), e.g.:

$$C_2H_5Br + \underset{\substack{\text{Sodium}\\ \text{methoxide}}}{CH_3ONa} \rightarrow \underset{\substack{\text{Methoxyethane}\\ \text{(ethyl methyl ether)}}}{C_2H_5.O.CH_3} + NaBr$$

15.6 The Hydroxyl Group

Monohydric alcohols, derived from the alkane series, have the general formula $C_nH_{2n+1}OH$. They are named by removing the '-e' from the corresponding alkane and replacing it with '-ol'. They can be classified as primary, secondary or tertiary alcohols, as in *Table 15.4*. Hence the characteristic groups are $-CH_2OH$ (primary), >CHOH (secondary) and ⋗COH (tertiary).

Table 15.4 Structure of alcohols

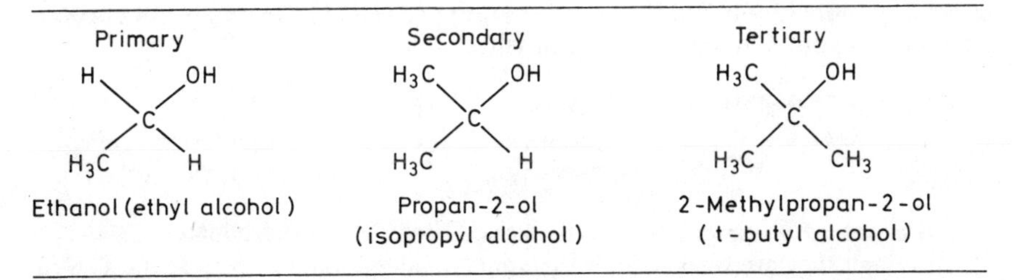

If the hydroxyl group is attached directly to the benzene ring, the compound is phenol, C_6H_5OH.

Alcohols can be prepared by the hydrolysis of halogenoalkanes (*see* p. 164). Phenol can be prepared from benzenediazonium chloride (p. 185).

15.6.1 REACTIONS AND PROPERTIES OF THE HYDROXYL GROUP

The lower alcohols, e.g., methanol, ethanol, etc., are liquids (because of hydrogen bonding) and are miscible with water. Phenol is a low melting point solid and is moderately soluble in water.

15.6.1.1 Acidity and Basicity

In principle, hydroxy-compounds can act as both acids and bases:

As an acid $\quad ROH \rightleftharpoons RO^- + H^+$

As a base: $\quad R\ddot{O}H + H^+ \rightleftharpoons ROH_2^+$

This acidity or basicity depends on the environment of the hydroxyl group. The pK_a values (p. 69) of phenol and methanol are 9.9 and ~16, respectively (cf. ethanoic acid, for example, 4.76). In the case of methanol, the electron pushing effect of the methyl group ($CH_3 \rightarrow OH$) means that the oxygen does not strongly attract the electrons in the O–H bond; hence proton release hardly occurs. Another way of looking at this is to say that the electron pushing effect of the alkyl group in alcohols promotes some lone pair donation by the oxygen atom, hence increasing its basicity.

In the case of phenol, it is a stronger acid than alcohols (but still weak) because (*a*) delocalization of non-bonding electrons on the oxygen into the nucleus makes oxygen electron poorer (p. 161); oxygen, therefore, attracts more

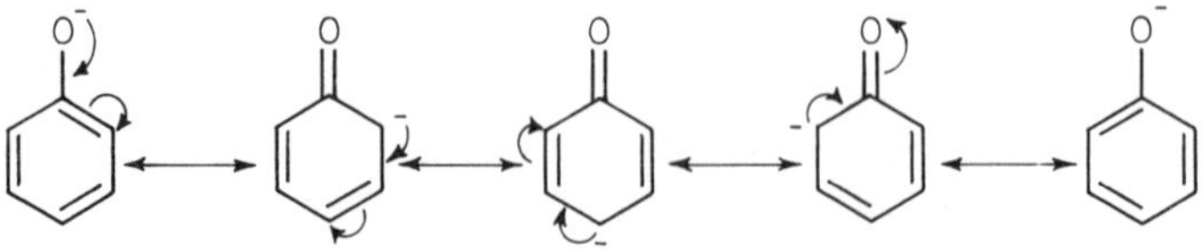

(27) (28) (29) (30) (31)

strongly the electrons in the O–H bond and (*b*) the resulting anion, after proton loss, is stabilized by resonance, as in structures (27)–(31). Hence phenol dissolves in sodium hydroxide solution to form salts,

$$C_6H_5OH + NaOH \rightarrow C_6H_5O^-Na^+ + H_2O$$

Sodium phenoxide

but it is not acidic enough to liberate carbon dioxide from carbonates.

Although they are weaker acids than phenol, alcohols do display acidic as well as basic properties – they react only slightly with alkalis but react with the very electropositive metals to form alkoxides, e.g.,

$$2CH_3OH + 2Na \rightarrow \underset{\text{Sodium methoxide}}{2CH_3O^-Na^+} + H_2$$

(Hence, sodium should not be used for drying alcohols.)

15.6.1.2 Oxidation

Primary, secondary and tertiary alcohols all behave differently on oxidation, and are therefore easily distinguished because of this. The usual laboratory oxidizing agents are acid or alkaline potassium manganate(VII) solution, acidified potassium dichromate(VI) solution or nitric acid.

Primary alcohols are oxidized to aldehydes, and then to acids:

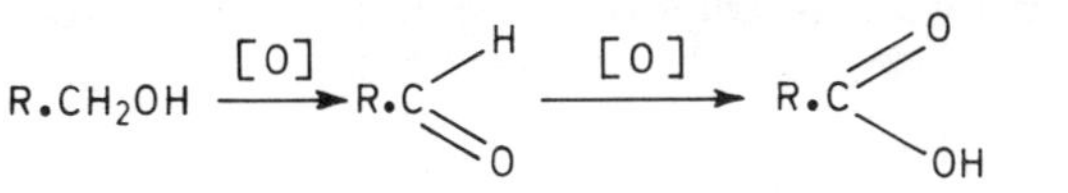

$$R.CH_2OH \xrightarrow{[O]} \underset{\text{An aldehyde}}{R.C(H)=O} \xrightarrow{[O]} \underset{\text{A carboxylic acid}}{R.C(=O)OH}$$

The aldehyde can be detected with 2,4-dinitrophenylhydrazine, and Tollens's or Schiff's reagent.

Secondary alcohols are oxidized to ketones (which can be detected by their reaction with 2,4-dinitrophenylhydrazine, but lack of reaction with Tollens's or Schiff's reagent):

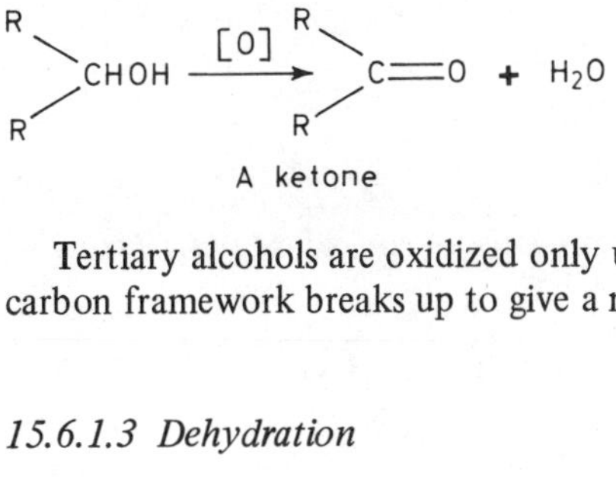

$$R_2CHOH \xrightarrow{[O]} \underset{\text{A ketone}}{R_2C{=}O} + H_2O$$

Tertiary alcohols are oxidized only under vigorous conditions, when the carbon framework breaks up to give a mixture of products.

15.6.1.3 Dehydration

Alcohols can be readily dehydrated to alkenes,

$$R.CH_2CH_2OH \xrightarrow{-H_2O} R.CH{=}CH_2$$

This can be done by passing ethanol, for example, over hot alumina (~350 °C) or by treating it with concentrated sulphuric acid (170 °C). In the latter case, the acid catalyst is necessary to convert the –OH into $-\overset{+}{O}H_2$, which is a good leaving group: *see Figure 15.3.* The ease of dehydration of alcohols is tertiary > secondary > primary. This is because the more alkyl groups there are attached to the carbonium ion, the more stable it is (p. 156).

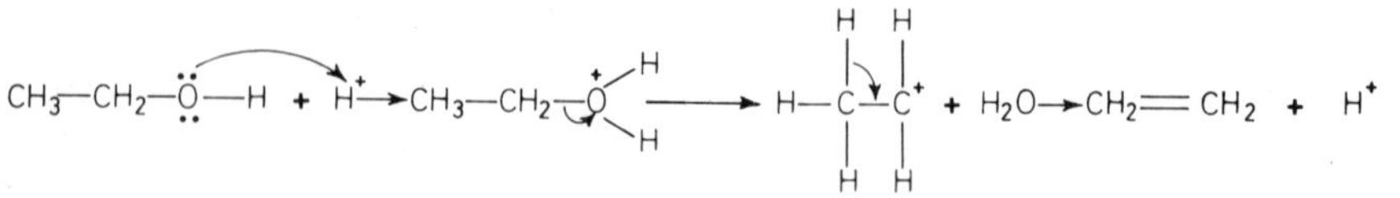

$$CH_3{-}CH_2{-}\ddot{O}{-}H + H^+ \rightarrow CH_3{-}CH_2{-}\overset{+}{O}H_2 \longrightarrow H{-}\underset{H}{\overset{H}{C}}{-}\underset{H}{\overset{H}{C^+}} + H_2O \rightarrow CH_2{=}CH_2 + H^+$$

Figure 15.3

It should be noted that at lower temperatures and by using excess alcohol, ethers are produced rather than alkenes, e.g.,

$$C_2H_5OH + H_2SO_4 \rightarrow C_2H_5.O.SO_2.OH + H_2O$$

$$C_2H_5.O.SO_2.OH + C_2H_5OH \rightarrow \underset{\text{(diethyl ether)}}{\underset{\text{Ethoxyethane}}{(C_2H_5)_2O}} + H_2SO_4$$

15.6.1.4 Halogenation

It has been seen that alcohols react with phosphorus halides (p. 163) and hydrogen halides on heating to form halogenoalkanes, e.g.,

$$C_2H_5OH + HBr \rightarrow C_2H_5Br + H_2O$$

In the above example, the reaction is carried out by heating the alcohol with sodium or potassium bromide and concentrated sulphuric acid (to generate hydrogen bromide). The mechanism involves protonation of the oxygen in the –OH group, which in turn makes the carbon of the alcohol more $\delta+$ and hence more easily attacked by Br^- (it also converts the –OH into $-\overset{+}{O}H_2$ which is a good leaving group):

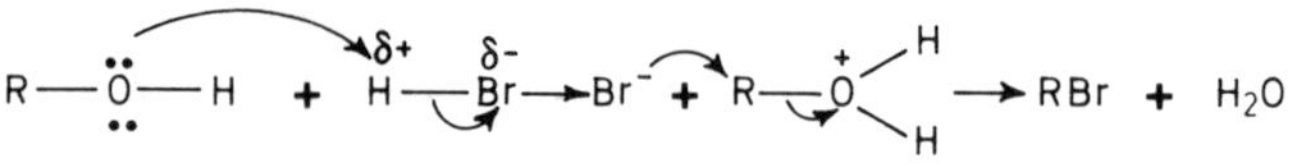

$$R{-}\ddot{O}{-}H + \overset{\delta+}{H}{-}\overset{\delta-}{Br} \rightarrow Br^- + R{-}\overset{+}{O}H_2 \longrightarrow RBr + H_2O$$

15.6.1.5 Ester Formation

When refluxed with carboxylic acids in the presence of an acid catalyst, alcohols form esters, e.g.,

$$\underset{\text{(acetic acid)}}{\underset{\text{Ethanoic acid}}{CH_3COOH}} + \underset{\text{Ethanol}}{C_2H_5OH} \overset{H^+}{\rightleftharpoons} \underset{\text{(ethyl acetate)}}{\underset{\text{Ethyl ethanoate}}{CH_3COOC_2H_5}} + H_2O$$

Phenol does not form esters with carboxylic acids, but does with acid chlorides and anhydrides (pp. 179 and 182).

15.6.1.6 Haloform Reaction

Alcohols containing the $CH_3.CHOH$ group give a yellow precipitate of triodomethane (iodoform) when warmed with iodine in dilute sodium hydroxide solution. Hence ethanol will give this reaction (and this test can be used to distinguish it from methanol):

$$C_2H_5OH + 4I_2 + 6NaOH \rightarrow CHI_3 + HCOONa + 5NaI + 5H_2O$$

The reaction is more general; with sodium chlorate(I) (hypochlorite) or sodium bromate(I) (hypobromite) these alcohols form trichloromethane (chloroform) or tribromomethane (bromoform) respectively.

15.6.1.7 Electrophilic Substitution With Phenol

Because phenol is electron rich in the 2-, 4-, and 6-positions of the ring (*see* p. 161), these positions are 'activated' towards electrophilic substitution, e.g., with bromine water at room temperature, a pale yellow precipitate of 2,4,6-tribromophenol (32) is formed.

OH

Br OH Br

$\xrightarrow{Br_2/H_2O}$ + 3 HBr

Br
(32)

15.7 Ethers

Ethers have the general structure R–O–R′ where R and R′ are alkyl or aryl groups (which may be the same or they may be different), as shown in *Table 15.5.*

Table 15.5 Ethers

Formula	*IUPAC name*	*Older name*
$CH_3.O.CH_3$	Methoxymethane	Dimethyl ether
$CH_3.O.C_2H_5$	Methoxyethane	Ethyl methyl ether
$C_2H_5.O.C_2H_5$	Ethoxyethane	Diethyl ether ('Ether')
$C_6H_5.O.C_2H_5$	Ethoxybenzene	Phenetole

Ethers may be prepared from the addition of concentrated sulphuric acid to alcohols (p. 168) or by the reaction of a halogenoalkane with an alkoxide (p. 165).

Ethers are generally more volatile than the corresponding alcohols since they contain no hydroxyl group for hydrogen bonding to occur, e.g., the two isomers $CH_3.O.CH_3$ and C_2H_5OH have boiling points of –24 °C and 78.5 °C, respectively. The lower ethers are highly flammable. Ethers are also much less reactive than alcohols because they do not have a slightly acidic hydrogen (i.e., the –OH hydrogen) which can be replaced by another group. Ethers, for example, can be dried over sodium wire (cf. alcohols, p. 167).

Like alcohols, ethers are weakly basic, dissolving in strong acids, e.g.,

$$C_2H_5.O.C_2H_5 + HBr \rightleftharpoons [C_2H_5.\underset{|\atop H}{O}.C_2H_5]^+ + Br^-$$

and forming stable complexes with Lewis acids, e.g., BF_3.

15.8 The Carbonyl Group

The carbonyl group has the general structure (33). Aldehydes have the general structure (34), where R=H or an alkyl or aryl group, and ketones have the general structure (35), where R = alkyl or aryl; the R groups may be the same or different.

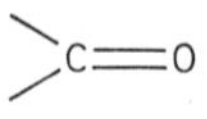

$$>C{=}O \qquad R{-}C(=O){-}H \qquad R_2C{=}O$$

(33) 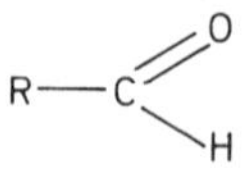 (34) 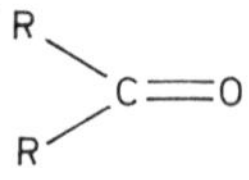 (35)

Aliphatic aldehydes are named after the hydrocarbon with the same number of carbon atoms, the end '-e' being replaced by 'al'; aromatic aldehydes are named as the hydrocarbon carbaldehyde as in *Table 15.6.* Aldehydes can be

Table 15.6 Aldehydes and ketones

Formula	*IUPAC name*	*Older name*
Aldehydes		
HCHO	Methanal	Formaldehyde
$CH_3.CHO$	Ethanal	Acetaldehyde
$CH_3.CH_2.CHO$	Propanal	Propionaldehyde
$C_6H_5.CHO$	Benzenecarbaldehyde	Benzaldehyde
Ketones (aliphatic)		
$CH_3.CO.CH_3$	Propanone	Acetone
$CH_3.CO.C_2H_5$	Butanone	Methyl ethyl ketone
$CH_3.CO.CH_2.CH_2.CH_3$	Pentan-2-one	Methyl n-propyl ketone
Ketones (aromatic)		
$C_6H_5.CO.CH_3$	Phenylethanone	Acetophenone
$C_6H_5.CO.C_6H_5$	Diphenylmethanone	Benzophenone

prepared by the oxidation of primary alcohols (p. 167). The oxidation must be carefully controlled to avoid further oxidation to a carboxylic acid (p. 167); acidified dichromate(VI) is the normal oxidizing agent. Aldehydes cannot be prepared by the reduction of carboxylic acids since further reduction to a primary alcohol occurs. However, aldehydes can be prepared by the catalytic hydrogenation of acid chlorides (the Rosenmund reaction):

$$R.COCl + H_2 \xrightarrow{Pd/BaSO_4} R.CHO + HCl$$

Aliphatic ketones are named by removing the '-e' from the corresponding alkane and replacing it with '-one', as in *Table 15.5.* Two common aromatic ketones are also shown in *Table 15.5.* Ketones are prepared by the oxidation of secondary alcohols (p. 160). Note that the Friedel–Crafts reaction (p. 162) gives aromatic ketones.

Simple aldehydes and ketones are colourless liquids (except methanal which is a gas at room temperature).

The carbonyl group contains a trigonal carbon atom, the carbon atom being

sp^2 hybridized (p. 21); hence, this part of aldehydes and ketones is planar, bond angles 120 degrees (*see* p. 21). Since the oxygen is more electronegative than the carbon, the carbonyl group is polarized, i.e., $>\overset{\delta+}{C}=\overset{\delta-}{O}$, and hence the $\delta+$ carbon undergoes nucleophilic attack (contrast the C=C bond which undergoes mainly electrophilic attack). The general order of reactivity of the carbonyl group (towards nucleophilic attack) is $H_2C{=}O > R.CHO > R_2C{=}O$, and this results from two factors. First, electron releasing groups attached to the carbonyl carbon atom will tend to reduce the $\delta+$ charge and hence make it less susceptible to nucleophilic attack. Secondly, bulky R groups will hinder the approach of a nucleophile (STERIC HINDRANCE).

15.8.1 SOME REACTIONS OF CARBONYL COMPOUNDS

15.8.1.1 Oxidation

Ketones are not readily oxidized, although vigorous oxidation gives a mixture of carboxylic acids containing fewer carbon atoms than the ketone. Aldehydes, however, are readily oxidized to carboxylic acids having the same number of carbon atoms as the aldehyde (p. 167). This can be done by using the usual strong oxidizing agents [acidified potassium manganate(VII), etc.] and also by using mild reagents such as Tollens's reagent and Fehling's solution (*see* below), i.e., aldehydes are strong reducing agents. In this way, aldehydes can be distinguished from ketones, and hence primary alcohols from secondary alcohols.

Hence aldehydes (but not ketones) reduce Fehling's solution [which can be thought of as containing copper(II) oxide as a complex]; a red precipitate of copper(I) oxide is formed on warming:

$$R.CHO + 2CuO \rightarrow R.COOH + Cu_2O\downarrow$$

Aldehydes (but not ketones) also reduce Tollens's reagent (made by adding a little sodium hydroxide solution to silver nitrate solution, followed by just sufficient ammonia solution to dissolve the precipitated silver oxide); on warming, a silver mirror is deposited on the walls of the container:

$$R.CHO + Ag_2O \rightarrow R.COOH + 2Ag$$

A further distinguishing test is that aldehydes restore the purple colour to Schiff's reagent (a dye, magenta, which has been decolourized by sulphur dioxide); this test is not given by ketones.

15.8.1.2 The Triodomethane (Iodoform) Reaction

This reaction is undergone by ethanal (acetaldehyde) and ketones containing the $-CO.CH_3$ group (also by alcohols which can be oxidized to carbonyl compounds containing the above group, e.g., ethanol, propan-2-ol, *see* p. 168), e.g., warming propanone with iodine and sodium hydroxide solution gives a yellow precipitate of triodomethane:

$$CH_3.CO.CH_3 + 3I_2 + 4NaOH \rightarrow CHI_3 + CH_3.COONa + 3NaI + 3H_2O$$

15.8.1.3 Polymerization

With concentrated sulphuric acid at room temperature, ethanal gives the ethanal trimer, $(CH_3CHO)_3$ (a liquid known as paraldehyde). Below 0 °C, the solid ethanal tetramer, $(CH_3CHO)_4$ (known as metaldehyde), is formed.

On standing in the presence of dilute aqueous alkali or sodium carbonate solution two molecules of ethanal join to form 3-hydroxybutanal (aldol); the reaction is usually referred to as the aldol condensation:

$$CH_3.CHO + CH_3.CHO \rightarrow CH_3.CHOH.CH_2.CHO$$

Methanal does not give this reaction. Ketones have little tendency to polymerize.

15.8.2 ADDITION REACTIONS OF CARBONYL COMPOUNDS

15.8.2.1 Reduction

This is essentially addition of hydrogen; aldehydes are reduced to primary alcohols, ketones to secondary alcohols:

$$R.CHO \xrightarrow{2[H]} R.CH_2OH$$

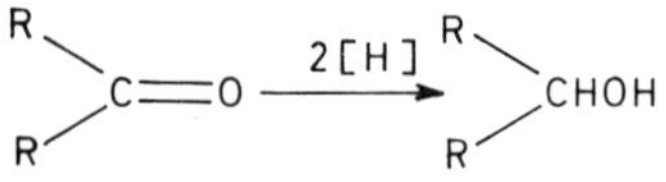

$$R_2C{=}O \xrightarrow{2[H]} R_2CHOH$$

Reagents which can be used for these reductions include: sodium and alcohol, lithium tetrahydridoaluminate (lithium aluminium hydride), sodium tetrahydridoborate (sodium borohydride), and hydrogen with a platinum or nickel catalyst. In the case of $LiAlH_4$ and $NaBH_4$, the essential feature of the mechanism is nucleophilic attack by AlH_4^- or BH_4^- on the carbonyl carbon.

15.8.2.2 Hydrogen Cyanide

This adds on to aldehydes and ketones to give hydroxynitriles (cyanohydrins). The reaction is base catalysed since this generates the CN^- nucleophile:

$$NaOH + HCN \rightarrow Na^+CN^- + H_2O$$

Then:

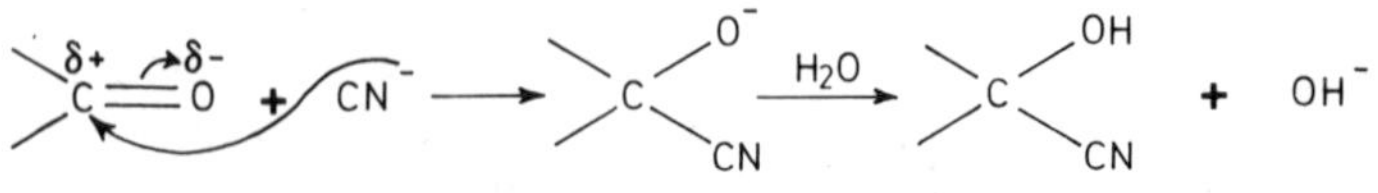

$$>\overset{\delta+}{C}{=}\overset{\delta-}{O} + CN^- \longrightarrow >C(O^-)(CN) \xrightarrow{H_2O} >C(OH)(CN) + OH^-$$

The principal use of this reaction is in synthesis, since hydrolysis of the nitrile (cyanide) to a carboxylic acid group under acidic conditions yields a hydroxycarboxylic acid, e.g., *Figure 15.4.*

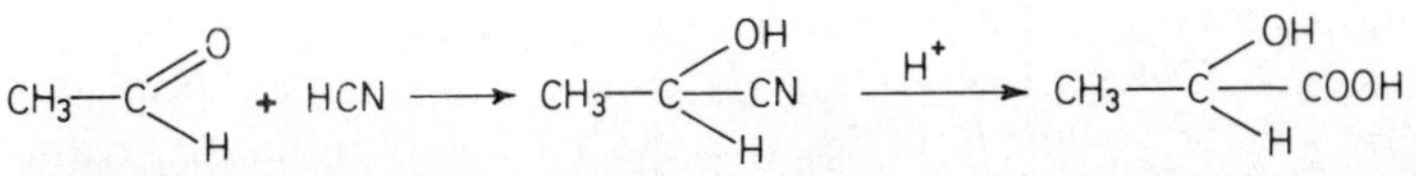

$$CH_3-C(=O)-H + HCN \longrightarrow CH_3-C(OH)(H)-CN \xrightarrow{H^+} CH_3-C(OH)(H)-COOH$$

Ethanal (acetaldehyde) 2-Hydroxypropanonitrile (acetaldehyde cyanohydrin) 2-Hydroxypropanoic acid (lactic acid)

Figure 15.4

15.8.2.3 Sodium Hydrogensulphite (Bisulphite)

When shaken with sodium hydrogensulphite solution, aldehydes and ketones give white crystalline addition compounds, e.g.,

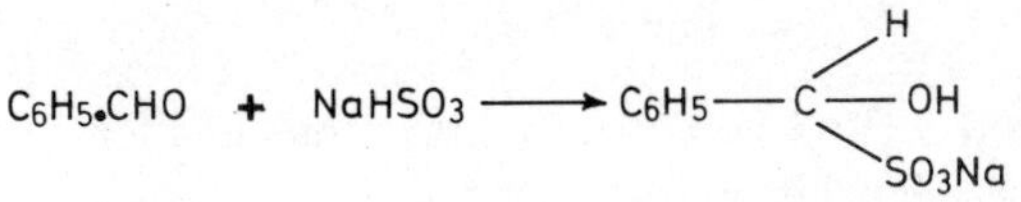

$$C_6H_5.CHO + NaHSO_3 \longrightarrow C_6H_5-C(H)(OH)-SO_3Na$$

Benzenecarbaldehyde sodium hydrogensulphite (benzaldehyde sodium bisulphite)

15.8.2.4 Alcohols

Addition of alcohols to aldehydes gives hemiacetals; reaction with a further molecule of alcohol gives acetals, e.g.,

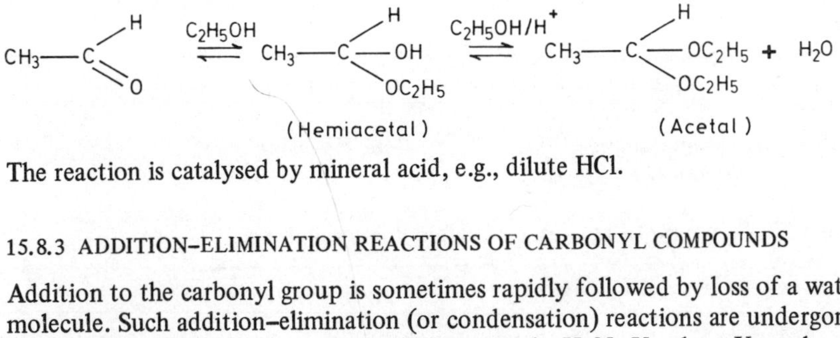

$$CH_3-C(H)=O \overset{C_2H_5OH}{\rightleftharpoons} CH_3-C(H)(OH)-OC_2H_5 \overset{C_2H_5OH/H^+}{\rightleftharpoons} CH_3-C(H)(OC_2H_5)-OC_2H_5 + H_2O$$

(Hemiacetal) (Acetal)

The reaction is catalysed by mineral acid, e.g., dilute HCl.

15.8.3 ADDITION–ELIMINATION REACTIONS OF CARBONYL COMPOUNDS

Addition to the carbonyl group is sometimes rapidly followed by loss of a water molecule. Such addition–elimination (or condensation) reactions are undergone particularly with substituted ammonia compounds, H_2N-X, where X can be a variety of groups (see below). The general mechanism is as shown in *Figure 15.5*.

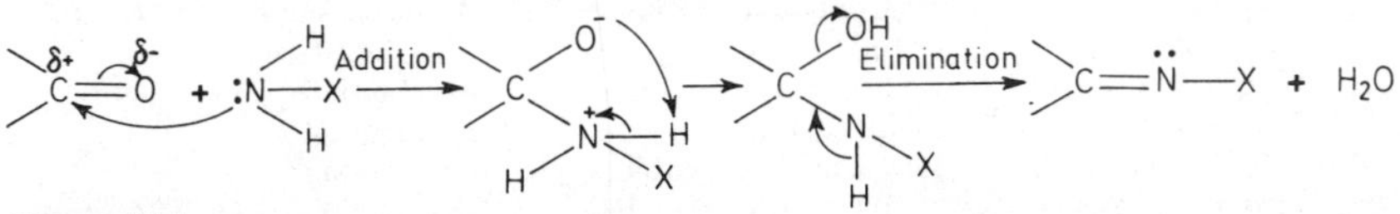

$$>C^{\delta+}=O^{\delta-} + :NH_2-X \xrightarrow{\text{Addition}} >C(O^-)-N^+H_2X \longrightarrow >C(OH)-NHX \xrightarrow{\text{Elimination}} >C=\ddot{N}-X + H_2O$$

Figure 15.5

Therefore, hydroxylamine, NH_2OH, reacts with aldehydes and ketones to form oximes, e.g.,

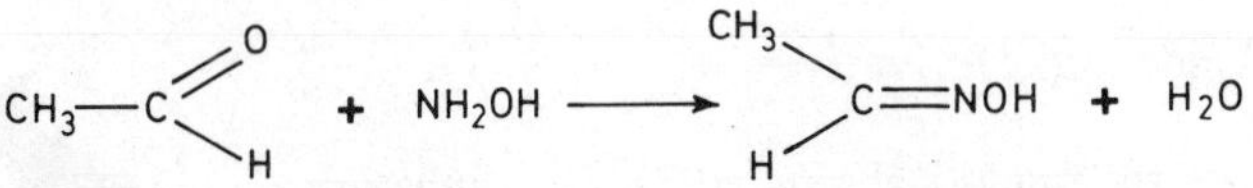

$$CH_3-C(=O)-H + NH_2OH \longrightarrow CH_3(H)C=NOH + H_2O$$

Ethanal oxime (acetaldehyde oxime)

With hydrazine, $NH_2.NH_2$, aldehydes and ketones give hydrazones. However, arylhydrazines such as phenylhydrazine, $C_6H_5.NH.NH_2$, and 2,4-dinitrophenylhydrazine, give solid crystalline phenylhydrazones and 2,4-dinitrophenylhydrazones respectively. The latter are yellow-orange precipitates, e.g., ethanal 2,4-dinitrophenylhydrazone (acetaldehyde 2,4-dinitrophenylhydrazone) (36).

$$CH_3{-}CHO + C_6H_3(NO_2)_2{-}HN.NH_2 \longrightarrow C_6H_3(NO_2)_2{-}HN.N{=}CH.CH_3 + H_2O$$

(36)

These derivatives have characteristic melting points and are therefore used for identifying the original carbonyl compound.

With aldehydes and ketones, semicarbazide gives semicarbazones, which are also crystalline and used for identification:

$$>C{=}O + NH_2.NH.CO.NH_2 \longrightarrow >C{=}N.NH.CO.NH_2 + H_2O$$

Semicarbazide — A semicarbazone

Amines react with carbonyl compounds to give substances known as Schiff's bases, e.g.,

$$>C{=}O + C_6H_5.NH_2 \longrightarrow >C{=}N.C_6H_5 + H_2O$$

Phenylamine (aniline) — A Schiff's base

Table 15.7 Summary of reactions of aldehydes and ketones

Test		*Aldehydes*	*Ketones*
(1)	Reduction	Form primary alcohols	Form secondary alcohols
(2)	HCN	Addition occurs	Addition occurs
(3)	$NaHSO_3$	Addition occurs	Addition occurs
(4)	NH_2X	Addition–elimination occurs	Addition–elimination occurs
(5)	Alcohols	Form acetals	Little reaction
(6)	Fehling's solution	Red precipitate	No reaction
(7)	Tollens's reagent	Silver mirror	No reaction
(8)	Schiff's reagent	Restores purple colour	No reaction
(9)	Triodomethane reaction	Formed with $CH_3.CHO$ only	Occurs with $-CO.CH_3$ group
(10)	Polymerization	Occurs	Little tendency to occur

15.9 Carboxylic Acids and Derivatives

Carboxylic acids contain the functional group (37). The aliphatic acids are named from the corresponding alkane, the ending '–e' being replaced by

'–oic acid', aromatic acids are named as the hydrocarbon carboxylic acid, as shown in *Table 15.8.*

Table 15.8 Carboxylic acids

HCOOH Methanoic acid (formic acid)	—C(=O)—O—H (37) The carboxyl group	CH_3•COOH Ethanoic acid (acetic acid)
CH_3—CH(CH_3)—COOH 2-Methylpropanoic acid (isobutyric acid)		C_6H_5COOH Benzenecarboxylic acid (benzoic acid)

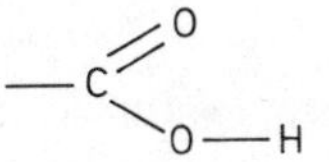

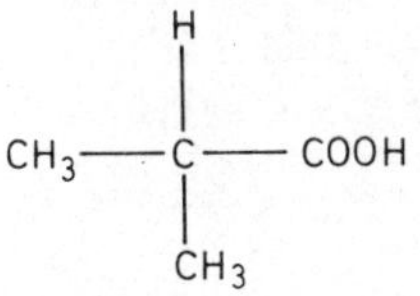

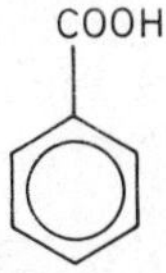

Carboxylic acids may be prepared by the oxidation of primary alcohols (p. 167), the oxidation of aldehydes (p. 167) or by the hydrolysis of nitriles (p. 182). Aromatic acids can be made by the oxidation of an alkyl side chain (p. 163).

The two simplest aliphatic acids, methanoic acid and ethanoic acid, have boiling points of 101 °C and 118 °C respectively; these relatively high values are attributed to hydrogen bonding (p. 25). Both of these acids, again because of hydrogen bonding, are soluble in water, but lengthening of the carbon chain does cause the acids to become less soluble. Benzenecarboxylic acid (m.p. 122 °C) is a white solid, which is sparingly soluble in cold water but fairly soluble in hot water.

An important feature of carboxylic acids is the lack of typical carbonyl properties. This can be partly explained by the resonance in the carboxyl group [structures (38) and (39)], which means that the carbonyl carbon atom is less

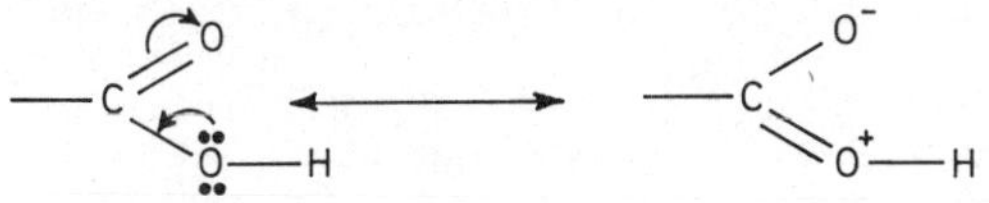

$$—C(=O)—\ddot{O}—H \quad (38) \longleftrightarrow \quad —C(—O^-)=O^+—H \quad (39)$$

electron deficient than in aldehydes and ketones. Furthermore in the liquid phase, carboxylic acids exist as dimers because of hydrogen bonding (p. 26), and therefore these hydrogen bonds must be broken (which requires energy) before carbonyl reactions can occur.

15.9.1 REACTIONS OF CARBOXYLIC ACIDS

15.9.1.1 Acidic Properties

Carboxylic acids are weak acids, the acidity being caused by the presence of the strongly electron attracting carbonyl group:

$$R.COOH + H_2O \rightleftharpoons R.COO^- + H_3O^+$$

Also, the resulting carboxylate ion is stabilized by resonance:

$$R-C(=O)O^- \longleftrightarrow R-C(O^-)=O$$

(40) (41)

The acidity of the acid depends on whether substituents in the R group are electron pushing (strengthens the O–H bond) or electron attracting (weakens the O–H bond). The examples in *Table 15.9* illustrate this point (*see* also p. 69). The pK_a of benzenecarboxylic acid is 4.20, making it a weaker acid than methanoic acid but stronger than ethanoic acid, i.e., the phenyl group has an overall electron donating effect compared with hydrogen.

Table 15.9 Structure and acidity in carboxylic acids

Structure	Name
H—CO_2H	Methanoic acid, pK_a 3.77 (formic acid)
$CH_3 \rightarrow CO_2H$	Ethanoic acid, pK_a 4.76 (acetic acid)
$CH_3 \rightarrow CH_2 \rightarrow CO_2H$	Propanoic acid, pK_a 4.88
$(CH_3 \rightarrow)_3C \rightarrow CO_2H$	2,2-Dimethylpropanoic acid, pK_a 5.05 (trimethylacetic acid)
$Cl \leftarrow CH_2.CO_2H$	Chloroethanoic acid, pK_a 2.86 (chloroacetic acid)
$(Cl \leftarrow)_2CH.CO_2H$	Dichloroethanoic acid, pK_a 1.29 (dichloroacetic acid)
$(Cl \leftarrow)_3C.CO_2H$	Trichloroethanoic acid, pK_a 0.65 (trichloroacetic acid)

Hence, carboxylic acids react with alkalis to give salts, and release carbon dioxide from carbonate solutions (used as a test for them), e.g.,

$$2CH_3.COOH + Na_2CO_3 \rightarrow \underset{\text{Sodium ethanoate (acetate)}}{2CH_3.COONa} + H_2O + CO_2$$

15.9.1.2 Ester Formation

On refluxing with alcohols, esters are produced (the reaction is acid catalysed):

$$R.COOH + R'OH \xrightleftharpoons{H^+/\text{reflux}} R.COOR' + H_2O$$

where R = H, alkyl or aryl and R′ = alkyl or aryl. Examples of esters include (42)–(44). The presence of the acid catalyst (e.g., concentrated H_2SO_4) also

$CH_3COOC_2H_5$	CH_3COOCH_3	$C_6H_5.COOC_6H_5$
Ethyl ethanoate (acetate)	Methyl ethanoate (acetate)	Phenyl benzenecarboxylate (benzoate)
(42)	(43)	(44)

absorbs water produced, hence displacing the equilibrium to the right. The mechanism is shown in *Figure 15.6.* Esters are readily hydrolysed, either by an acid catalysed mechanism (reverse of above) or a base catalysed mechanism

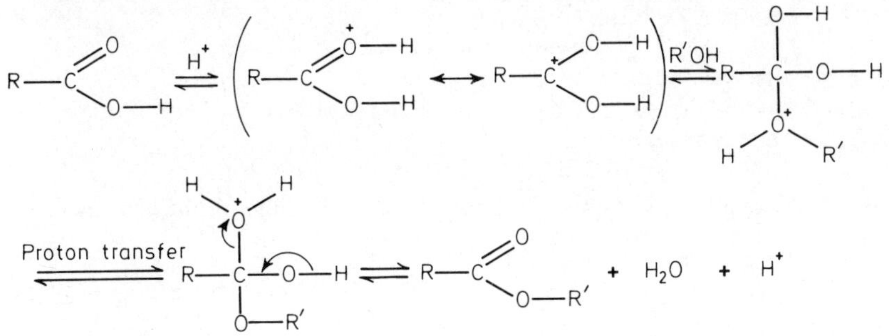

Figure 15.6

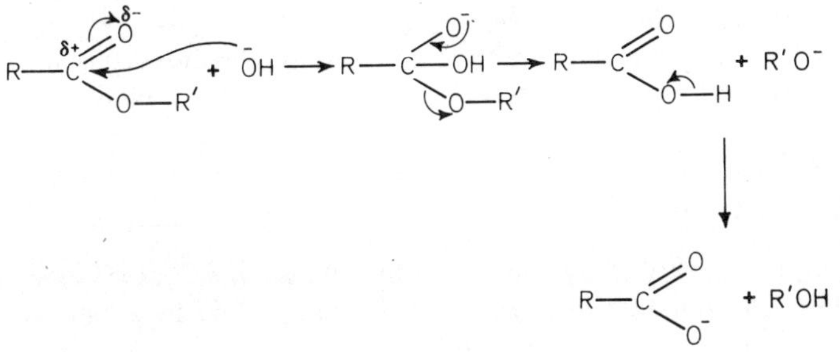

Figure 15.7

shown in *Figure 15.7.* Note that the R.COO⁻ ion is left rather than the R′O⁻ ion, since the former can be stabilized by resonance (p. 176).

15.9.1.3 Reduction

Carboxylic acids are reduced by lithium tetrahydridoaluminate, $LiAlH_4$, to primary alcohols:

$$R.COOH \xrightarrow{LiAlH_4} R.CH_2OH$$

15.9.1.4 Halogenation

Halogenation of the alkyl groups occurs with chlorine or bromine in the presence of light or a catalyst (e.g., iodine or red phosphorus):

$$CH_3.CO_2H \xrightarrow{Cl_2} CH_2Cl.CO_2H \xrightarrow{Cl_2} CHCl_2.CO_2H \xrightarrow{Cl_2} CCl_3.CO_2H$$

The hydroxyl group can also be replaced by halogens using phosphorus pentachloride or sulphur dichloride oxide (thionyl chloride), $SOCl_2$, e.g.,

$$PCl_5 + CH_3.COOH \rightarrow CH_3.COCl + HCl + POCl_3$$

Ethanoyl chloride (acetyl chloride)

$$C_6H_5.COOH + SOCl_2 \rightarrow C_6H_5.COCl + SO_2 + HCl$$

Benzenecarbonyl chloride (benzoyl chloride)

15.9.1.5 Alkane Formation

Sodium salts of carboxylic acids when heated with soda lime produce alkanes (p. 153).

15.9.2 ACID (ACYL) CHLORIDES

The acid or acyl chlorides have the functional group –COCl, shown in structure (45), and are named by removing the '-ic acid' from the acid name and replacing it with '-yl chloride'. They are prepared by the halogenation of carboxylic acids (*see* above), and are generally fuming liquids. Their properties are as follows.

15.9.2.1 Hydrolysis

Aliphatic acid chlorides are very easily hydrolysed by cold water since they are very susceptible to nucleophilic attack (because of the carbonyl polarization and because the chlorine is a good leaving group, as Cl^-):

$$R.COCl + H_2O \rightarrow R.COOH + HCl$$

The mechanism is shown in *Figure 15.8.* Aromatic acid chlorides hydrolyse more slowly because the carbonyl carbon atom is not so electron deficient (because of possible orbital overlap with the ring).

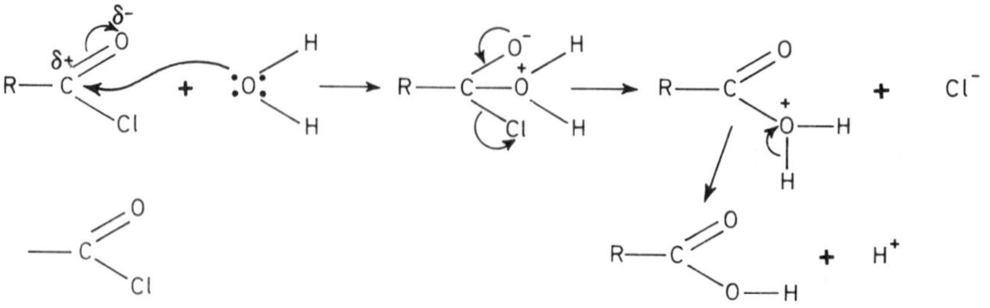

(45)

Figure 15.8

15.9.2.2 With Alcohols

A similar reaction to hydrolysis occurs, resulting in the formation of esters, e.g.,

$$CH_3.COCl + C_2H_5OH \rightarrow CH_3.COOC_2H_5 + HCl$$

$$C_6H_5COCl + C_6H_5OH + NaOH \rightarrow C_6H_5.COOC_6H_5 + H_2O + NaCl$$

In the case of the latter reaction, the base is used to generate the stronger nucleophile $C_6H_5O^-$ (phenoxide ion). The process of replacing a hydrogen atom by a $CH_3.CO–$ or $C_6H_5.CO–$ group is known as acylation.

15.9.2.3 With Ammonia

This reaction produces amides; for example, ethanoyl (acetyl) chloride reacts vigorously at room temperature with ammonia to form ethanamide (acetamide):

$$CH_3.COCl + 2NH_3 \rightarrow CH_3.CO.NH_2 + NH_4Cl$$

15.9.2.4 With Amines

With primary and secondary amines, *N*- and *N,N*-disubstituted amides, respectively, are formed, e.g.,

$$CH_3.COCl + (CH_3)_2NH \rightarrow \underset{\substack{N,N\text{-Dimethylethanamide} \\ (N,N\text{-dimethylacetamide})}}{CH_3.CO.N(CH_3)_2} + HCl$$

$$C_6H_5.COCl + \underset{\substack{\text{Phenylamine} \\ \text{(aniline)}}}{C_6H_5NH_2} + NaOH \rightarrow \underset{\substack{N\text{-Phenylbenzenecarboxamide} \\ \text{(benzanilide)}}}{C_6H_5.CO.NH.C_6H_5} + H_2O + NaCl$$

The mechanisms for these reactions (and with ammonia) are similar to that for hydrolysis.

15.9.2.5 With Sodium Salts of Acids

Anhydrides are produced (p. 181).

15.9.2.6 With Benzene

Ketones are produced (*see* Friedel–Crafts reaction, p. 162).

15.9.3 AMIDES

Amides have the functional group $–CONH_2$ shown in structure (46), and are named by removing the '-oic acid' from the acid name and replacing it with

'-amide'. They may be prepared from the reaction between acid chlorides and ammonia (*see* above) or by heating the ammonium salt of a carboxylic acid:

$$R.COONH_4 \xrightarrow{\text{heat}} R.CO.NH_2 + H_2O$$

Most amides are solids; the lower members are water soluble. Amides are only very weakly basic because the lone pair of electrons on the nitrogen is rendered less available for co-ordination because of resonance:

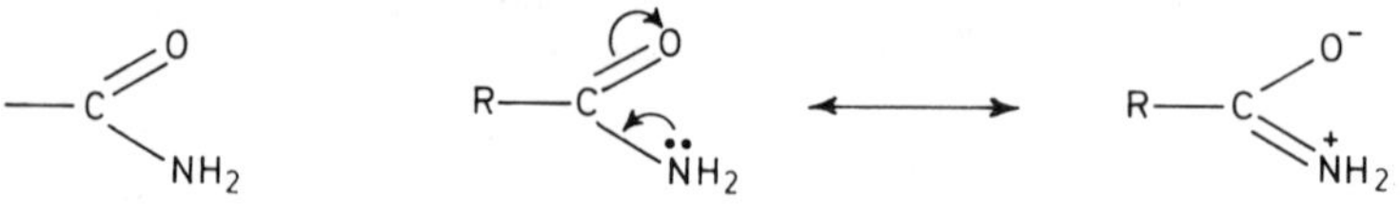

(46)

Their properties are as follows.

15.9.3.1 Hydrolysis

Amides can be hydrolysed slowly by boiling water but more rapidly in the presence of a mineral acid or alkali, e.g.,

$$R.CO.NH_2 + H_2O \rightarrow R.COOH + NH_3$$

In the presence of base, the mechanism is shown in *Figure 15.9*.

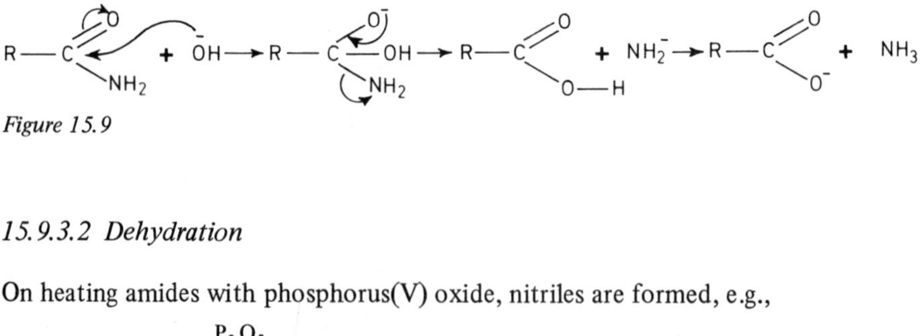

Figure 15.9

15.9.3.2 Dehydration

On heating amides with phosphorus(V) oxide, nitriles are formed, e.g.,

$$\underset{\substack{\text{Ethanamide}\\\text{(acetamide)}}}{CH_3.CO.NH_2} \xrightarrow{P_2O_5} \underset{\substack{\text{Ethanonitrile}\\\text{(acetonitrile)}\\\text{(methyl cyanide)}}}{CH_3.CN} + H_2O$$

15.9.3.3 Reduction

Amides are reduced to amines using sodium and ethanol, lithium tetrahydrido-aluminate, etc., e.g.,

$$CH_3.CO.NH_2 \xrightarrow{4[H]} \underset{\text{Ethylamine}}{CH_3.CH_2.NH_2} + H_2O$$

15.9.3.4 Hofmann Degradation

On heating with bromine and potassium hydroxide solution, amides give amines containing one less carbon atom, e.g.,

$$R.CO.NH_2 \xrightarrow[\text{heat}]{Br_2/KOH} R.NH_2 \text{ (primary amine)}$$

15.9.4 ANHYDRIDES

Anhydrides have the structure $(R.CO)_2O$ [as in structure (47)] where R = alkyl or aryl. They are named as, for example, $(CH_3.CO)_2O$, ethanoic anhydride (acetic anhydride), and can be regarded as the result of loss of water between two carboxyl groups.

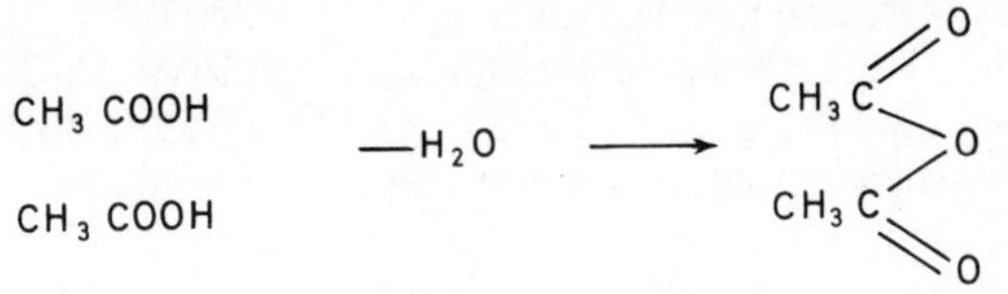

$$\begin{matrix} CH_3COOH \\ CH_3COOH \end{matrix} \quad -H_2O \longrightarrow (CH_3CO)_2O \quad (47)$$

They are prepared by the reaction between acid chlorides and sodium salts of acids, e.g.,

$$CH_3.COCl + CH_3.COONa \rightarrow (CH_3.CO)_2O + NaCl$$

The mechanism is shown in *Figure 15.10.*

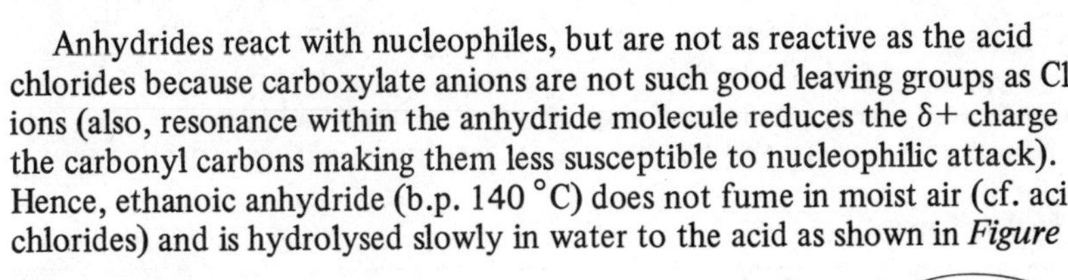

$$CH_3-C^{\delta+}(=O^{\delta-})Cl + CH_3COO^- \longrightarrow CH_3-C(O^-)(Cl)-O-C(=O)-CH_3 \longrightarrow (CH_3CO)_2O \; (47) + Cl^-$$

Figure 15.10

Anhydrides react with nucleophiles, but are not as reactive as the acid chlorides because carboxylate anions are not such good leaving groups as Cl^- ions (also, resonance within the anhydride molecule reduces the $\delta+$ charge on the carbonyl carbons making them less susceptible to nucleophilic attack). Hence, ethanoic anhydride (b.p. 140 °C) does not fume in moist air (cf. acid chlorides) and is hydrolysed slowly in water to the acid as shown in *Figure 15.11.*

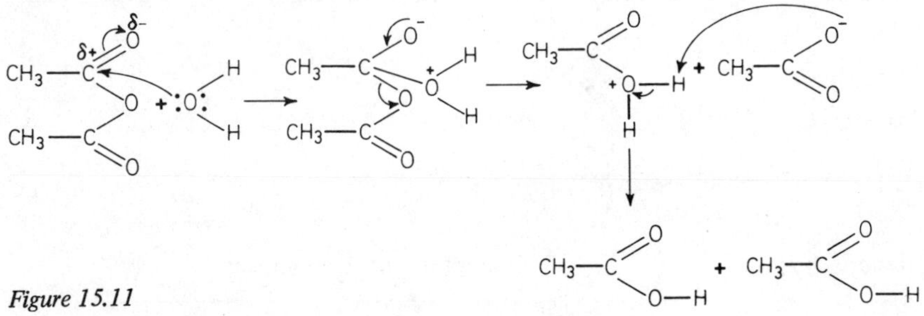

$$(CH_3CO)_2O + H_2O \longrightarrow CH_3-C(O^-)(O^+H_2)-O-COCH_3 \longrightarrow CH_3COO^+H_2 + CH_3COO^- \longrightarrow CH_3-C(=O)-O-H + CH_3-C(=O)-O-H$$

Figure 15.11

With alcohols and phenols, esters are produced, the mechanism being similar to that for hydrolysis, e.g.,

$$(CH_3.CO)_2O + C_2H_5OH \rightarrow CH_3.COOC_2H_5 + CH_3.COOH$$

With ammonia, amides are formed, e.g.,

$$(R.CO)_2O + 2NH_3 \rightarrow R.CO.NH_2 + R.COONH_4$$

Again, the mechanism is similar to that for hydrolysis.

15.9.5 NITRILES

Nitriles contain the functional group —C≡N. They are named from the parent hydrocarbon, e.g., $CH_3.CN$ is ethanonitrile (acetonitrile or methyl cyanide), $CH_3.CH_2.CN$ is propanonitrile (propionitrile or ethyl cyanide), and so on.

Nitriles can be prepared by refluxing halogenoalkanes with sodium or potassium cyanide solution (p. 165) or by the dehydration of amides (p. 180).

Ethanonitrile is a colourless liquid (b.p. 81 °C) and is soluble in water. Higher members of the series are only slightly soluble in water. Benzenecarbonitrile (benzonitrile), $C_6H_5.C{\equiv}N$, has a boiling point of 191 °C.

Nitriles are readily hydrolysed by mineral acids or alkalis:

$$R.CN \xrightarrow{H_2O} R.CO.NH_2 \xrightarrow{H_2O} R.COONH_4 \text{ (or } R.COOH + NH_3\text{)}$$

Acid hydrolysis gives the carboxylic acid (*via* the amide), whilst alkaline hydrolysis gives the alkali metal salt of the same carboxylic acid.

Nitriles are reduced to primary amines using sodium and ethanol:

$$R.CN \xrightarrow{4[H]} R.CH_2.NH_2$$

The following (which can be used for both aliphatic and aromatic nitriles) offers a way of introducing an extra carbon atom into a chain, e.g.,

$$R.OH \xrightarrow{HI} RI \xrightarrow{CN^-} R.CN \xrightarrow{\text{hydrolysis}} R.COOH \xrightarrow{LiAlH_4} R.CH_2OH$$

15.10 Amines

Compounds containing the amino-group are called amines and they can be classified as primary, secondary and tertiary amines, as shown in *Table 15.10.* The R

Table 15.10 Types of amine

$R—\ddot{N}H_2$ Primary amine	$R_2\ddot{N}H$ (R\NH/R) Secondary amine
$R_3N{:}$ Tertiary amine	$R_4N^+\ X^-$ Quaternary ammonium salt

groups can be alkyl or aryl (or any combination of these). The quaternary ammonium salt is a substituted ammonium salt where all four hydrogens have been replaced by organic groups, and X^- can be Cl^-, I^-, etc.

Amines are named according to the groups present, e.g., *Table 15.11*.

Table 15.11 Nomenclature of some amines

$CH_3.NH_2$ Methylamine	$CH_3.CH_2.NH_2$ Ethylamine	$C_6H_5.NH_2$ Phenylamine (aniline)
$(CH_3)_2NH$ Dimethylamine	$(CH_3)_3N$ Trimethylamine	$C_6H_5.NH.CH_3$ *N*-Methylphenylamine (*N*-methylaniline)

Primary amines may be prepared by the reduction of nitriles (p. 182), amides (p. 180) and nitro-compounds. The latter method is the most useful for preparing aromatic amines, e.g.,

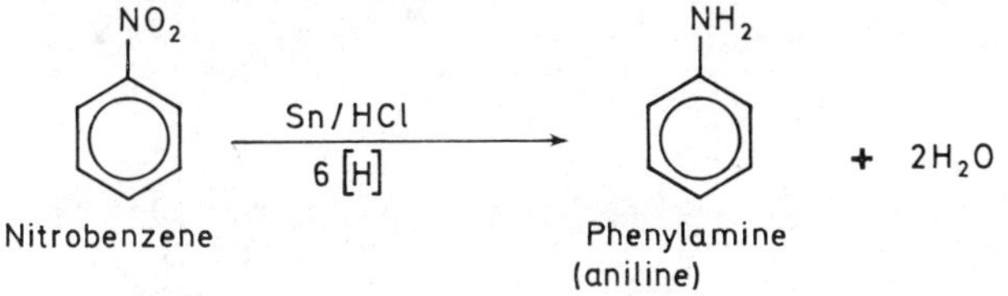

$$C_6H_5NO_2 \xrightarrow[6[H]]{Sn/HCl} C_6H_5NH_2 + 2H_2O$$

Nitrobenzene → Phenylamine (aniline)

After this reduction, the reaction mixture is made alkaline (by adding slaked lime) and the phenylamine, which is volatile in steam, is steam distilled (*see* pp. 57 and 144) from the reaction mixture.

Primary, secondary and tertiary amines (and quaternary ammonium salts) can be prepared by heating halogenoalkanes (alkyl halides) with ammonia in a sealed vessel, e.g.,

$$C_2H_5I + NH_3 \rightarrow C_2H_5.\overset{+}{N}H_3I^- \quad [C_2H_5.NH_2 + HI]$$
$$C_2H_5I + C_2H_5.NH_2 \rightarrow (C_2H_5)_2\overset{+}{N}H_2I^- \quad [(C_2H_5)_2NH + HI]$$
$$C_2H_5I + (C_2H_5)_2NH \rightarrow (C_2H_5)_3\overset{+}{N}HI^- \quad [(C_2H_5)_3N + HI]$$
$$C_2H_5I + (C_2H_5)_3N \rightarrow (C_2H_5)_4N^+I^-$$

Tetraethylammonium iodide

The above reactions involve nucleophilic attack as shown, for example:

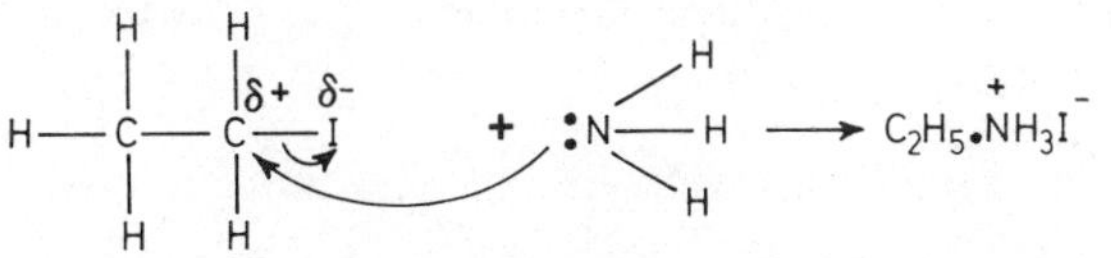

$$CH_3-CH_2^{\delta+}-I^{\delta-} + :NH_3 \longrightarrow C_2H_5.\overset{+}{N}H_3I^-$$

The lower aliphatic amines are gases (higher members are liquids) and they possess a strong 'fishy' smell. Aromatic amines are high boiling liquids or solids, e.g., phenylamine, b.p. 184 °C. Lower amines are soluble in water because of the polarization in the N–H bond (*see* also p. 27).

15.10.1 REACTIONS OF AMINES

15.10.1.1 Basic Properties

Amines exhibit basic properties because of the lone pair on the nitrogen atom; hence they can form salts with acids, e.g.,

$$C_6H_5.NH_2 \underset{OH^-}{\overset{HCl}{\rightleftharpoons}} C_6H_5.\overset{+}{N}H_3Cl^-$$

Phenylammonium chloride
(aniline hydrochloride)

Note that the base can be released again using alkali.

The relative base strengths of amines depend on the availability of the lone pair for co-ordination, and hence on substituents present, e.g.,

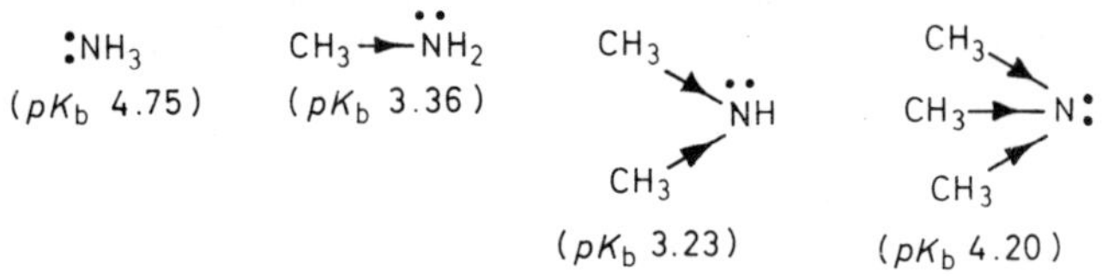

$:NH_3$ (pK_b 4.75) $CH_3 \rightarrow \ddot{N}H_2$ (pK_b 3.36) $(CH_3)_2\ddot{N}H$ (pK_b 3.23) $(CH_3)_3N:$ (pK_b 4.20)

As the pK_b values indicate (page 69), the base strengths increase as alkyl groups are introduced into the ammonia; however, the tertiary amine has an unexpectedly high value of pK_b, i.e., its base strength is surprisingly low. This is because when a base is dissolved in water, the following equilibrium is established (*see* also p. 69):

$$R_3N: + H_2O \rightleftharpoons R_3\overset{+}{N}H + OH^-$$

The base strength not only depends on electron pair availability but also on how stable the resultant cation can become by solvation – this latter effect is greater for a secondary amine than for a tertiary amine, as shown in *Figure 15.12.*

(a) (b)

Figure 15.12 Structure (a) is more stable than structure (b) because the former contains more hydrogen bonding

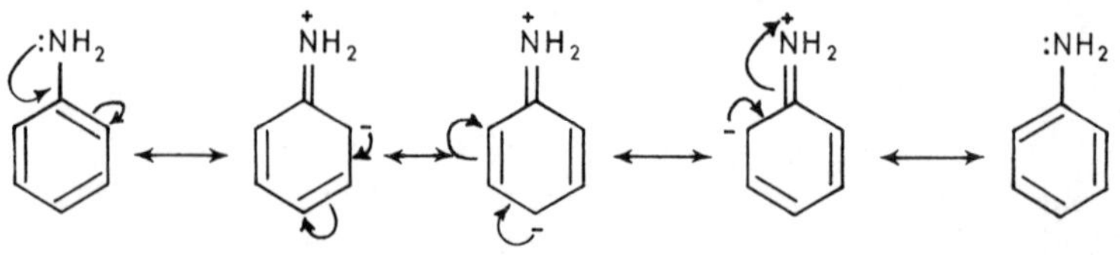

Figure 15.13

Aromatic amines are weaker bases than aliphatic amines (phenylamine, pK_b 9.38) because the lone pair can become delocalized in the ring, as in *Figure 15.13.*

15.10.1.2 Acylations and Alkylations

Acylation of amines is described on p. 179. Alkylations of primary amines to form secondary and tertiary amines and quaternary ammonium salts are described above.

15.10.1.3 With Nitrous Acid

With nitrous acid and a primary aliphatic amine in the cold, a primary alcohol is formed and nitrogen is evolved, e.g.,

$$C_2H_5.NH_2 + HNO_2 \rightarrow C_2H_5OH + N_2 + H_2O$$

With an aromatic primary amine, the addition of aqueous sodium nitrite to phenylamine, for example, in excess dilute hydrochloric acid at a temperature of between 5 °C and 10 °C yields the unstable compound benzenediazonium chloride:

$$C_6H_5.NH_2 + HNO_2 + HCl \rightarrow C_6H_5.N_2^+Cl^- + 2H_2O$$

This reaction is characteristic of all primary aromatic amines where the amino-group is attached directly to the ring. If the temperature is allowed to rise, the diazonium salt decomposes to phenol.

15.11 Properties of Diazonium Salts

Reactions of diazonium salts can be divided into (1) those where the nitrogen is evolved and (2) those where the nitrogen is retained.

15.11.1 REACTIONS INVOLVING REPLACEMENT OF NITROGEN

15.11.1.1 By Hydroxyl Group

On heating a diazonium salt solution, phenol is produced:

$$C_6H_5.N_2^+Cl^- + H_2O \rightarrow C_6H_5OH + N_2 + HCl$$

15.11.1.2 By Chlorine or Bromine

On warming a diazonium salt solution with either a copper(I) halide in its halogen acid or with powdered copper, replacement by a halogen occurs, e.g.,

$$C_6H_5.N_2^+Cl^- \xrightarrow{Cu} C_6H_5Cl + N_2$$

15.11.1.3 By Iodine

On warming a diazonium salt solution with potassium iodide solution, replacement by iodine occurs:

$$C_6H_5.N_2^+Cl^- + KI \rightarrow C_6H_5I + N_2 + KCl$$

15.11.2 REACTIONS WHERE THE NITROGEN IS RETAINED

Diazonium salts 'couple' with phenols and aromatic amines to give highly coloured azo compounds, many of which are used commercially. For example, with an alkaline solution of phenol, the diazonium salt gives a bright orange precipitate of (4-hydroxyphenyl)azobenzene (*p*-hydroxyazobenzene):

$$C_6H_5{-}N_2^+Cl^- + C_6H_5{-}OH \xrightarrow{NaOH} C_6H_5{-}N{=}N{-}C_6H_4{-}OH + NaCl + H_2O$$

If naphthalen-2-ol (β-naphthol) is used instead, a bright red precipitate is formed.

An example of coupling with a tertiary aromatic amine is the reaction between *N,N*-dimethylphenylamine (*N,N*-dimethylaniline) and the diazonium salt:

$$C_6H_5{-}N_2^+Cl^- + C_6H_5{-}N(CH_3)_2 \rightarrow C_6H_5{-}N{=}N{-}C_6H_4{-}N(CH_3)_2 + HCl$$

The product is 4-(phenylazo)-*N,N*-dimethylphenylamine (*p*-dimethylaminoazobenzene) (yellow).

16

SOME BIOLOGICALLY IMPORTANT ORGANIC COMPOUNDS

16.1 Amino-Acids

When proteins are broken up by hydrolysis, α-amino-acids are formed (in which the amino-group is attached to the carbon atom adjacent to the carboxyl group, called the α-carbon atom). Two of the simplest amino-acids are (1) and (2).

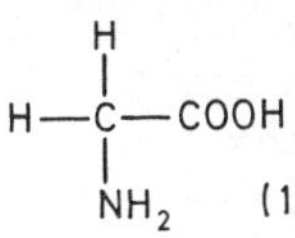

$$H-\underset{NH_2}{\overset{H}{C}}-COOH \quad (1) \qquad\qquad CH_3-\underset{NH_2}{\overset{H}{C}}-COOH \quad (2)$$

Aminoethanoic acid (glycine) 2-Aminopropanoic acid (alanine)

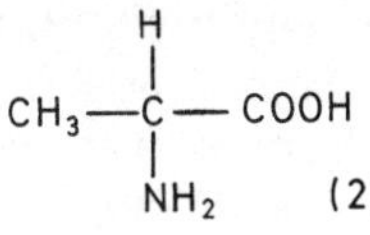

Note that 2-aminopropanoic acid contains an asymmetric carbon atom and so exhibits optical isomerism (*see* Chapter 14, p. 148).

The α-amino-acids can be prepared in two ways:

(1) By treating α-halogeno-acids with concentrated ammonia solution:

$$CH_3COOH + Cl_2 \rightarrow CH_2Cl.COOH + HCl$$

$$CH_2Cl.COOH + 2NH_3 \rightarrow CH_2(NH_2).COOH + NH_4Cl$$

(2) From aldehydes, *via* their hydroxynitriles (cyanohydrins), using hydrogen cyanide in the presence of ammonia:

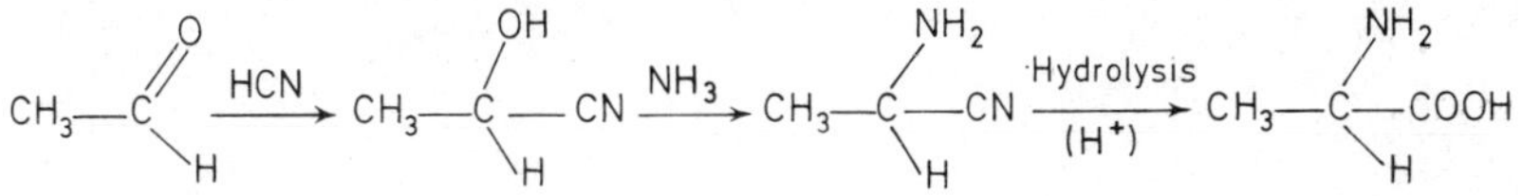

$$CH_3-C(=O)H \xrightarrow{HCN} CH_3-C(OH)(H)-CN \xrightarrow{NH_3} CH_3-C(NH_2)(H)-CN \xrightarrow[(H^+)]{\text{Hydrolysis}} CH_3-C(NH_2)(H)-COOH$$

As these molecules contain both the amino and carboxyl groups, they show the reactions of both functional groups (*see* amines and carboxylic acids; Chapter 15). Also, because one group is basic and the other is acidic, they exist as INNER SALTS or ZWITTERIONS, e.g.,

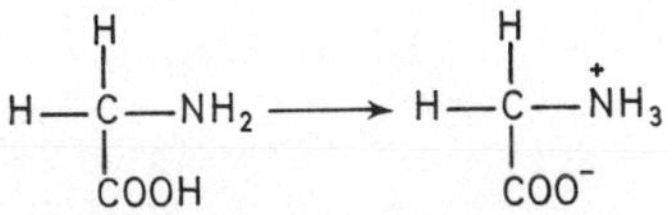

$$H-\underset{COOH}{\overset{H}{C}}-NH_2 \longrightarrow H-\underset{COO^-}{\overset{H}{C}}-\overset{+}{N}H_3$$

Hence, their properties are 'salt-like', having high melting points, and tending to be soluble in water rather than organic, non-polar solvents.

16.2 Polypeptides and Proteins

As stated earlier, when proteins are hydrolysed, amino-acids are produced, which shows that the protein contains the POLYPEPTIDE chain (3). Note that if water is added to the part of the chain in brackets, an amino-acid results. The –CO.NH– link is called the PEPTIDE LINK.

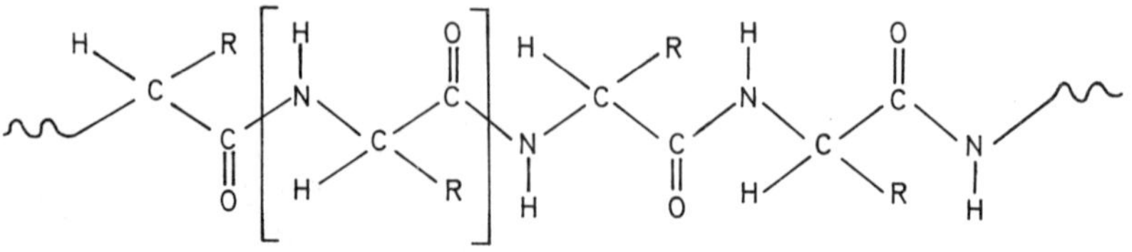

(3)

Proteins are naturally occurring polypeptides, having very long chains, and occur in all living cells, a particular protein being peculiar to each type of cell. Relative molecular masses of proteins vary widely, e.g., insulin, approximately 6000; egg albumen, approximately 40 000; haemoglobin, approximately 67 000. The long protein chains are often helical in form because hydrogen bonding between the –NH and >C=O groups on the same chain (intramolecular forces – *see* p. 25) bend it into this configuration.

16.3 Carbohydrates

Carbohydrates, consisting of carbon, hydrogen and oxygen only, are constituents of plants and animals, and include the sugars and macromolecular substances such as starch and cellulose. Carbohydrates have the general formula $C_x(H_2O)_y$. They can be classified into two main groups: sugars and non-sugars. Sugars, which are generally sweet-tasting, crystalline and water-soluble, can be further divided into mono-, di-, tri-saccharides, etc. The non-sugars are polysaccharides, i.e., they are macromolecular structures built up from many monosaccharide units joined together by the loss of water. Hence, polysaccharides can be broken down on hydrolysis to form monosaccharides.

16.3.1 SUGARS

The two best known monosaccharides are glucose and fructose, both of molecular formula $C_6H_{12}O_6$, i.e., they are isomeric. They are referred to as HEXOSES because they both contain six carbon atoms. Both glucose and fructose occur in fruits and honey.

Glucose is manufactured by the hydrolysis of starch in the presence of a mineral acid (at high pressure and temperature):

$$\underset{\text{(Starch)}}{(C_6H_{10}O_5)_n} + nH_2O \rightarrow \underset{\text{(Glucose)}}{n(C_6H_{12}O_6)}$$

Glucose has the structure (4); note that it contains four asymmetric carbon atoms, and so the molecule is optically active. Because of its structure, glucose

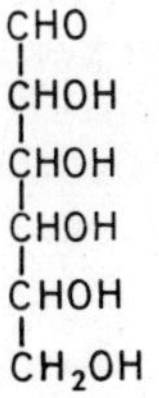

CHO
CHOH
CHOH
CHOH
CHOH
CH_2OH

(4)

CHOH
CHOH O
CHOH
CHOH
CH
CH_2OH

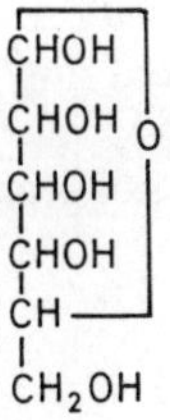

(5)

shows the properties of both alcohols and aldehydes (chapter 15), i.e., it reacts with ethanoic (acetic) anhydride to form a pentaethanoate (pentaacetate) indicating the presence of five hydroxyl groups; it reduces Fehling's solution and Tollens's reagent (p. 171) and thus behaves like an aldehyde (sugars which exhibit these reducing properties are called REDUCING SUGARS). However, there is further chemical evidence that the 'open-chain' structure of glucose (4) is not quite correct, but that it exists also as a ring structure (5), formed as a result of internal hemiacetal formation (*see* p. 173).

For fructose (obtained by the hydrolysis of maltose or sucrose – *see* below), there is also evidence for both open and closed chain formulae (because of internal hemiacetal formation). Note that there are three asymmetric carbon atoms in structure (6). Fructose reduces Fehling's solution and Tollens's reagent.

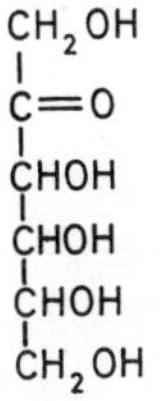

CH_2OH
C=O
CHOH
CHOH
CHOH
CH_2OH

(6)

CH_2OH
HOC
CHOH
CHOH O
CHOH
CH_2

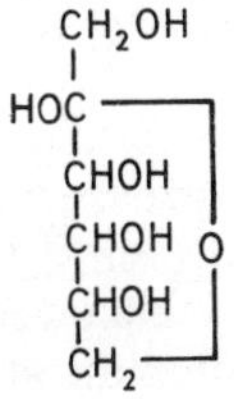

(7)

Although ketones do not normally show these reactions, an α-hydroxyketone (a ketone having a hydroxyl group on a carbon atom adjacent to the carbonyl carbon) docs give these reactions.

The molecular formula of the disaccharides is $C_{12}H_{22}O_{11}$. On hydrolysis, one molecule of disaccharide and one molecule of water form two molecules of formula $C_6H_{12}O_6$, and these two molecules can be of the same hexose or they can be different ones:

$$C_{12}H_{22}O_{11} + H_2O \rightarrow 2C_6H_{12}O_6$$

Two common disaccharides are sucrose and maltose.

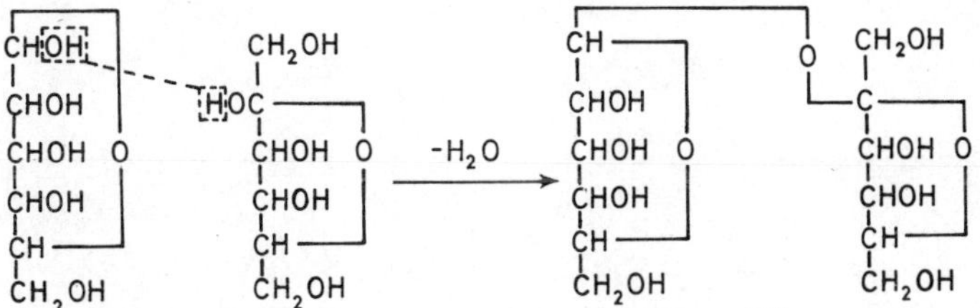

CHOH | CH_2OH
CHOH | HOC
CHOH O | CHOH O
CHOH | CHOH
CH | CH
CH_2OH | CH_2OH

(8)

$-H_2O$ →

CH | O | CH_2OH
CHOH | | C
CHOH O | | CHOH O
CHOH | | CHOH
CH | | CH
CH_2OH | | CH_2OH

(9)

Sucrose (found in sugar cane and sugar beet) can be considered as being made from glucose and a five-membered ring form of fructose, as shown in structures (8) and (9). Since both reducing groups are involved in the linkage, sucrose is a non-reducing sugar, i.e., it does not react with Fehling's solution or Tollens's reagent.

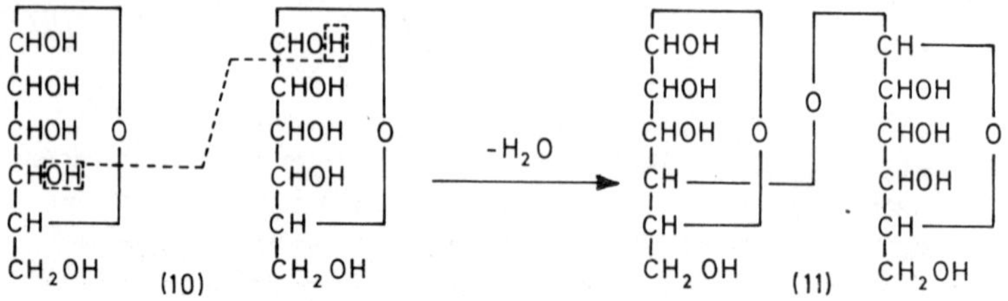

```
CHOH                CHOH                         CHOH             CH
CHOH                CHOH                         CHOH        O    CHOH
CHOH   O            CHOH   O       -H2O          CHOH   O         CHOH   O
CHOH                CHOH          ------>        CH               CHOH
CH                  CH                           CH               CH
CH2OH     (10)      CH2OH                        CH2OH     (11)   CH2OH
```

Maltose (present in malt) contains two glucose units linked together. It can be seen that the top >CHOH group of the left hand glucose unit (10) is the aldehyde group when in open chain form, i.e., the reducing group. Hence maltose is a reducing sugar, reacting with Fehling's solution and Tollens's reagent.

16.3.2 NON-SUGARS

The two most important polysaccharides are starch and cellulose, both consisting of glucose units.

Cellulose, the main constituent of the cell walls of plants, has the formula $(C_6H_{10}O_5)_n$. On hydrolysis with acids, it gives glucose. The glucose units are linked *via* their aldehyde carbon atoms [structure (12)], making cellulose non-reducing, i.e., no reaction with Fehling's solution or Tollens's reagent. Commercially, cellulose is very important in the manufacture of paper.

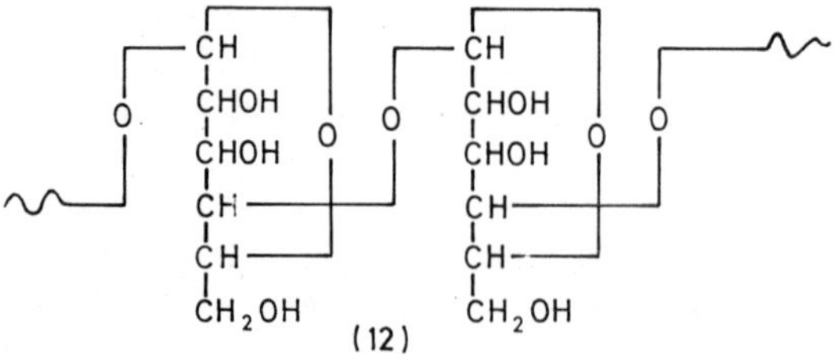

```
        CH                  CH
   O    CHOH       O   O    CHOH       O   O
        CHOH                CHOH
~       CH                  CH
        CH                  CH
        CH2OH     (12)      CH2OH
```

Starch is present in all green plants, wheat, barley, maize, rice, etc. It also has a molecular formula $(C_6H_{10}O_5)_n$, and gives glucose on hydrolysis. Starch is non-reducing, having a similar structure to that of cellulose. Starch gives a deep blue colour with iodine (*see* p. 121).

17

SYNTHETIC MACROMOLECULES

17.1 Polymers

High polymers, or MACROMOLECULES, consist of very large molecules which are built up by the repetition of small chemical units; this 'building-up' process is called POLYMERIZATION. The repeating unit is usually the same, or is very nearly the same, as the starting material or MONOMER from which the polymer is made.

Polymerizations can be classified into ADDITION POLYMERIZATIONS and CONDENSATION POLYMERIZATIONS. Addition polymerizations involve CHAIN reactions, the chain carriers being free radicals or ions; condensation polymerizations involve the elimination of small molecules, such as water.

17.2 Addition Polymerization

Ethene, and many of its derivatives, are polymerized by this method. Examples are shown in *Table 17.1.*

Table 17.1 Addition polymerizations

$$nCH_2{=\!=}CH_2 \longrightarrow {+}CH_2{-}CH_2{+}_n$$ poly(ethene) (Polythene)

$$nCH_2{=\!=}CHCl \longrightarrow {+}CH_2{-}\underset{\displaystyle Cl}{\underset{|}{CH}}{+}_n$$ poly(chloroethene)(polyvinyl chloride)

$$nCH_2{=\!=}CH{\cdot}CN \longrightarrow {+}CH_2{-}\underset{\displaystyle CN}{\underset{|}{CH}}{+}_n$$ poly(propenonitrile) (polyacrylonitrile)

$$nCH_2{=\!=}CH{\cdot}C_6H_5 \longrightarrow {+}CH_2{-}\underset{\displaystyle C_6H_5}{\underset{|}{CH}}{+}_n$$ poly(phenylethene) (polystyrene)

$$nCF_2{=\!=}CF_2 \longrightarrow {+}CF_2{-}CF_2{+}_n$$ poly(tetrafluoroethene)(P.T.F.E. or Teflon)

Free radical polymerization is probably the most common type of addition polymerization. The free radicals can be generated by the thermal or photochemical decomposition of compounds such as peroxides, e.g., di(benzoyl) peroxide [benzoyl peroxide (1)]:

$$C_6H_5-C(=O)-O-O-C(=O)-C_6H_5 \;(1) \xrightarrow[\text{or u.v.}]{\text{Heat}} 2C_6H_5-C(=O)-O\cdot$$

The free radical (called R· for convenience) can now initiate a free radical chain reaction:

$$R\cdot \; \dot{C}H_2=CH_2 \longrightarrow R-CH_2-\dot{C}H_2 \xrightarrow{CH_2=CH_2} R-CH_2-CH_2-CH_2-\dot{C}H_2$$

and so on. The process can be stopped by either COMBINATION:

$$\sim\sim\dot{C}H_2 + \dot{C}H_2\sim\sim \longrightarrow \sim\sim CH_2-CH_2\sim\sim$$

or by DISPROPORTIONATION (transfer of a hydrogen atom from one radical to another):

$$\sim\sim\dot{C}H_2 + \dot{C}H_2-CH(H)\sim\sim \longrightarrow \sim\sim CH_3 + CH_2=CH\sim\sim$$

Other types of addition polymerization include cationic polymerization and anionic polymerization.

Ziegler–Natta type catalysts [e.g., $(C_2H_5)_3Al$ with $TiCl_3$ or $TiCl_4$] are particularly useful in industry. At one time, poly(ethene) was only produced by free radical polymerization at pressures between 1000 and 3000 atmospheres and temperatures up to 250 °C. The above catalysts now enable ethene to be polymerized in solution at normal temperatures and pressures. Their mechanisms are not fully understood.

It should be noted that when alkenes of general formula $CH_2=CHX$ (where X can be C_6H_5 etc.) are polymerized, the X group can be situated on the same side of the chain, on alternate sides of the chain, or randomly situated on either side of the chain, and these are then known as isotactic, syndiotactic and atactic polymers respectively. The type of polymer made governs its properties, since if the substituents are arranged regularly, as in isotactic polymers, then chains can approach each other more closely (hence allowing interchain forces to operate) and the crystallinity of the polymer sample is increased (*see* also *Figure 4.10*). Interchain forces (between regular chains) can be increased by the presence of polar groups, causing dipole–dipole attractions or hydrogen bonding (*see* below).

Some uses of addition polymers include: poly(ethene) for food packaging (particularly as a foil or film) and containers generally, insulation, piping; poly(chloroethene) (P.V.C.) for packaging, piping, protective clothing; poly(propenonitrile) for synthetic fibres, such as orlon, etc; poly(phenylethene) for insulation, mouldings, coatings and sheets, foams for heat insulation; poly(tetrafluoroethene) for non-stick surfaces and insulation.

17.3 Condensation Polymerization

Two important classes of condensation polymer are the polyesters, e.g., Terylene, and the polyamides, e.g., Nylon.

Terylene can be prepared by heating ethane-1,2-diol (ethylene glycol) with dimethyl benzene-1,4-dicarboxylate (dimethyl terephthalate); the dimethyl ester of benzene-1,4-dicarboxylic acid (terephthalic acid) is used instead of the acid since the ester is easier to purify. The small molecule eliminated (*see Figure 17.1*) in this polymerization is methanol, CH_3OH, and is distilled off as it is formed. Polymers of high relative molecular mass are obtained.

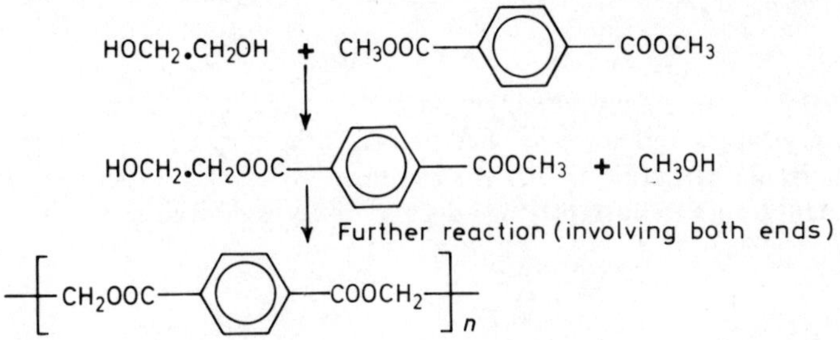

$HOCH_2.CH_2OH$ + $CH_3OOC-C_6H_4-COOCH_3$

↓

$HOCH_2.CH_2OOC-C_6H_4-COOCH_3$ + CH_3OH

↓ Further reaction (involving both ends)

$\left[CH_2OOC-C_6H_4-COOCH_2 \right]_n$

Figure 17.1

The word NYLON is used generally to denote synthetic fibre-forming polyamides, and the different nylons are distinguished by using a numbering system which indicates the number of carbon atoms in the repeating unit. Nylon 6.6, for example, is formed by heating hexanedioic acid (adipic acid) with hexane-1,6-diamine (hexamethylenediamine), resulting in the elimination of water as shown in *Figure 17.2.* Nylon 6, however, is made from cyclohexanol, as shown in *Figure 17.3.*

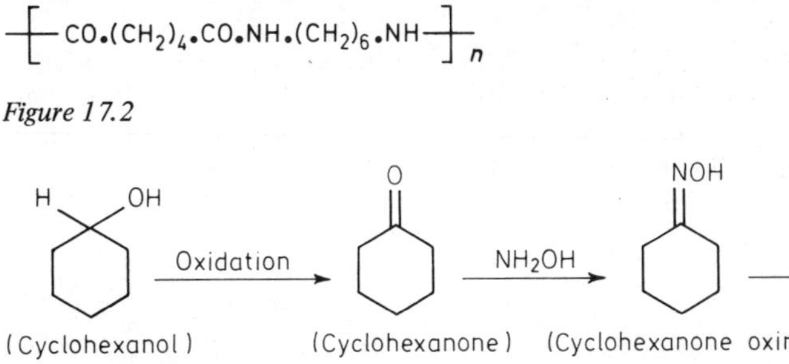

$HOOC.(CH_2)_4.COOH$ + $NH_2.(CH_2)_6.NH_2$

↓

$HOOC.(CH_2)_4.CO.NH.(CH_2)_6.NH_2$

↓ Further reaction (involving both ends)

$\left[CO.(CH_2)_4.CO.NH.(CH_2)_6.NH \right]_n$

Figure 17.2

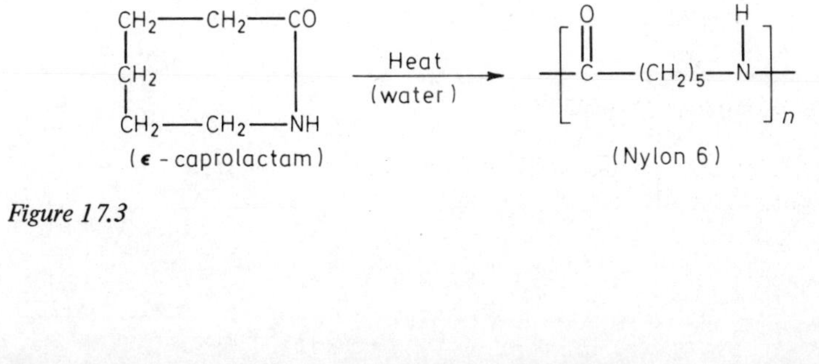

(Cyclohexanol) —Oxidation→ (Cyclohexanone) —NH_2OH→ (Cyclohexanone oxime) —H^+→

CH_2-CH_2-CO / CH_2 / CH_2-CH_2-NH (ε - caprolactam) —Heat (water)→ $\left[C(=O)-(CH_2)_5-N(H) \right]_n$ (Nylon 6)

Figure 17.3

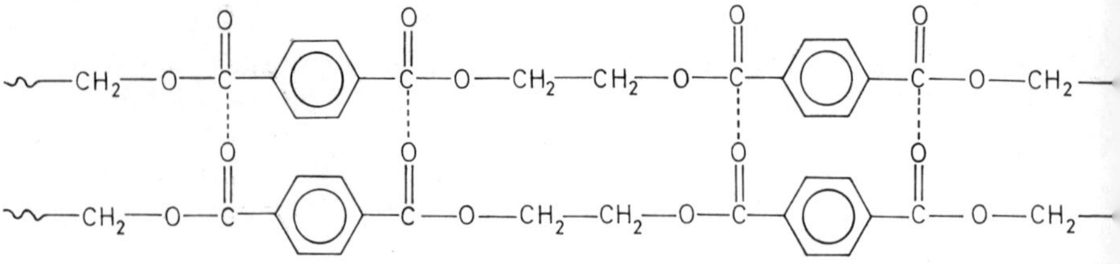

(2)

Terylene and the nylons are extremely important for producing fibres. Fibres are often highly crystalline and, as mentioned before, if the polymer molecules have a regular molecular structure so that chains can approach each other closely (enabling interchain forces to operate), high degrees of crystallinity can result.

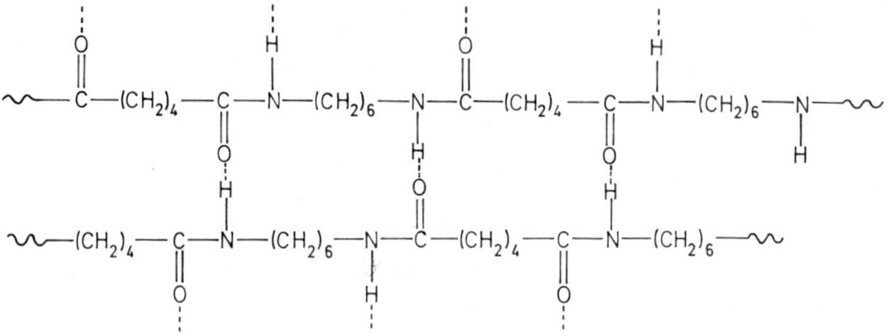

(3)

In the case of Terylene, dipole–dipole attractions can occur between chains (which help to maintain an ordered structure) as in structure (2), whereas in the case of the polyamide chains, hydrogen bonding can occur as in structure (3).

17.3.1 CROSS LINKING

Polymer chains may be joined together across the chains (do not confuse with combination; *see* before) by CROSS-LINKS to give a CROSS-LINKED polymer. This cross-linking process can reduce the solubility of the polymer, and can also hinder chain movement. When rubber is VULCANIZED, hydrocarbon chains are joined or linked by –S–S– cross links.

INDEX

Absolute zero, 28
Absorption coefficient, 50
Acetaldehyde (*see* Ethanal)
Acetals, 173
Acetamide (*see* Ethanamide)
Acetic acid (*see* Ethanoic acid)
Acetone (*see* Propanone)
Acetophenone (*see* Phenylethanone)
Acetyl chloride (*see* Ethanoyl chloride)
Acetylene (*see* Ethyne)
Acid anhydrides, 179, 181–182
Acid-base titrations, 73–75
Acid chlorides, 162, 178–179, 181
Acids, 65
    carboxylic, 26, 174–178
    conjugate, 65
    dissociation constant of, 66, 69
    strengths of, 68, 69, 176
Activated complex, 83
Activation energy, 83
Acylation, 162, 179
Addition polymerisation, 191–192
Adipic acid (*see* Hexanedioic acid)
Alcohols, 150, 165–169, 171, 172, 173, 176, 177, 179, 182
Aldehydes, 167, 170–174
Aldol, 172
    condensation, 172
Alkali metals, 38, 94–98
Alkaline earth metals, 94–98
Alkanes, 152–154
Alkanesulphonic acids, 153, 154
Alkenes, 154–158, 167, 168
Alkoxides, 165, 169
Alkyl group, 12, 151, 152
    halides (*see* Halogenoalkanes)
Alkylation, 162
Alkynes, 158
    structure of, 22
Allotropy, 102, 109, 116
Alpha particles, 17, 18, 19
Aluminium, 98
    extraction, 80, 101
    compounds, 11, 84, 90, 98–101, 117, 161
Amides, 179–181, 182
Amines, 181, 182–185
Amino-acids, 148, 187–188
Ammonia, 20, 21, 63, 112, 113, 114, 115
Amorphous solids, 39–40
Amphiprotic nature, 70
Analysis of compounds, 144–146
Anhydrides, 179, 181–182
Aniline (*see* Phenylamine)
Anion, 7, 13
Anode, 77, 78
Anodising, 100
Antifreeze, 158
Argon, 125
Aromatic hydrocarbons, 158–163
Arrhenius equation, 84
Association, 26, 55
Astatine, 120
Asymmetric carbon atom, 148, 187
Atactic polymers, 192
Atom, 1–5, 14–19
Atomic mass, relative, 14
    number, 1, 5, 6, 88
    radius, 6, 87–88
    spectra, 1, 4–5
    volume, 88
Aufbau principle, 2
Avogadro's hypothesis, 30
Azeotropic mixtures, 57

Balmer series, 5
Bands, 36
Barium and compounds, 94–98
Bases, 65
    conjugate, 65
    dissociation constant of, 69
    strengths of, 69, 70, 184
Bauxite, 101
Becquerel, 16
Benzal chloride (*see* (Dichloromethyl)benzene)
Benzaldehyde (*see* Benzenecarbaldehyde)
Benzanilide (*see* *N*-Phenylbenzenecarboxamide)
Benzene, 13, 22, 23, 43, 66, 158–163
Benzenecarbaldehyde, 170, 173
Benzenecarbonitrile, 182
Benzenecarbonyl chloride, 178
Benzenecarboxylic acid, 163, 175
Benzenediazonium chloride, 185, 186
Benzene-1,4-dicarboxylic acid, 193
Benzene hexachloride (*see* 1,2,3,4,5,6-Hexachlorocyclohexane)
Benzoic acid (*see* Benzenecarboxylic acid)
Benzophenone (*see* Diphenylmethanone)

Benzoyl chloride (*see* Benzenecarbonyl chloride)
Benzoyl peroxide (*see* Di(benzoyl) peroxide)
Benzyl chloride (*see* (Chloromethyl) benzene)
Beryllium and compounds, 16, 20, 90–91, 94–98
Beta rays, 17, 18
Binary solutions, 50–57
Blast furnace, 136–138
Body-centred cubic structure, 37–38
Boiling point elevation, 51–52
Bond angles, 10, 20, 21, 22, 23
energies, 10, 44
lengths, 22
Bonding molecular orbital, 8
Born–Haber cycle, 44–46
Boron and compounds, 2, 9, 20, 21, 98–99, 169
Boyle's Law, 28, 29
Bragg equation, 36
Bromine and compounds, 120–125, 155, 156, 169, 185
Bromoethane, 156
Brønsted and Lowry (acids and bases), 65
Brown ring test for nitrates, 110
Brownian movement, 34
Buffer solutions, 70–72
Butane, 152
Butanone, 170
Butenedioic acids, 147
Butenedioic anhydride, 162
t-Butyl alcohol (*see* 2-Methylpropan-2-ol)

Caesium and compounds, 94–97
Calcium and compounds, 62, 94–97, 143, 145, 158
Calomel electrode, 75, 77
Camphor, 53
ϵ-Caprolactam, 193
Carbanion, 151
Carbohydrates, 188–190
Carbon and compounds, 101–108
Carbonium ion, 151
stabilisation of, 156
Carbonyl chloride, 105
Carbonyl group, 170–174
Carboxylic acids, 174–178
Catalysts, 63, 84, 115, 119, 128, 162, 176, 177
Catenation, 102–103, 115
Cation, 7, 13
Cell diagram, 77
Cells, 76–79
Cellulose, 53, 190
Chadwick, 16
Chain reaction, 191–192
Charles–Gay–Lussac law, 28
Chemical equilibria, 60–63
Chlorides, chemical properties and periodicity of, 90
Chlorination, 85, 153, 162
Chlorine and compounds, 14, 85, 120–125, 139, 153, 155, 162, 177, 185
Chloroacetic acid (*see* Chloroethanoic acid)
Chlorobenzene, 163, 164
Chloroethanoic acid, 69, 176
Chloroethene, 191
Chloroform (*see* Trichloromethane)
Chloromethane, 21, 85, 153
(Chloromethyl)benzene, 162
2-Chloro-2-methylpropane, 86
Chromatography, 144
Chromium, 127, 133
compounds, 133–134
Cis-isomer, 147
Close packing, 36–38
Cobalt, 127, 139
compounds, 139–140
Colligative properties, 50–55
Common ion effect, 65
Complex ions, 129, 130
Condensation polymerisation, 192–194
Conductivity of metals, 36
Constant boiling point mixture, 57
Contact process, 63, 119
Convergence limit, 5, 87
Co-ordinate bond, 11, 129
Co-ordination number, 130
Copper, 127, 140
compounds, 140–141
Cottrell's boiling point apparatus, 52
Covalent bond, 8–12
Cracking, 154
Critical pressure, 34
temperature, 34
Cross linking, 194
Cryoscopic constant, 52
Crystal structure, 34–38
Cubic close packing, 37
Cyanohydrins, 172
Cyclohexane, 159
Cyclohexanol, 193
Cyclohexanone, 193
Cyclohexene, 159

d-orbital, 2, 6, 9, 127–130
Dalton's law of partial pressures, 29–30
Daniell cell, 77–78
Dative bond, 11, 129
De Broglie, 16
Degree of dissociation, 66
Dehydration of alcohols, 154, 167, 168
of amides, 180, 182
Dehydrohalogenation, 154, 165
Delocalisation, 23, 36, 128

Depression of freezing point, 52–53
Diagonal relationship, 90–91
Diamagnetism, 111
Diamond, 38–39, 102
Diastereoisomers, 149
Diazonium salts, 185–186
Di(benzoyl) peroxide, 191
1,2-Dibromoethane, 155
Dichloroethanoic acid, 69, 176, 177
Dichloromethane, 85, 153
(Dichloromethyl)benzene, 162
Diethyl ether (*see* Ethoxyethane)
Diffusion, 32
Dimerization, 26
Dimethyl benzene-1,4-dicarboxylate, 193
N,N-Dimethylethanamide, 179
Dimethyl ether (*see* Methoxymethane)
N,N-Dimethylphenylamine, 186
1,3-Dinitrobenzene, 160
Dinitrogen oxide, 109, 114
Dinitrogen pentoxide, 81, 82
Dinitrogen tetraoxide, 110, 111
2,4-Dinitrophenylhydrazine, 174
Diphenylmethanone, 170
Dipole–dipole attractions, 24
Disaccharides, 189
Disproportionation, 105
Dissociate, 26, 55
Dissociation constant, 66
Distillation, 56, 143–144
Distribution coefficient, 55
Drying agents, 143

Ebullioscopic constant, 51
Einstein's equation, 16
Electrochemical cells, 76–79
Electrochemistry, 76–80
Electromagnetic spectrum, 17
Electron, 1, 14–15
Electron affinity, 7, 87
Electron pair repulsion theory, 20–22
Electronic configuration, 2, 3
Electrophile, 151
Electrophilic substitution, 160–162
Electropositivity, 7, 94
Electrovalent bond, 12
Elevation of boiling point, 51–52
Elimination, 164, 165
Ellingham diagram, 138
Emission spectrum, 4–5
Empirical formula, 144–145
Enantiomer, 148
Endothermic, 7, 42, 83
Energetics, 41–48
Enthalpy change, 41
Enthalpy of atomisation, 42, 43
  of combustion, 42
  of formation, 42
  of hydrogenation, 43
  of solution, 46
Entropy, 47, 48
Equilibria, chemical, 60–63
  ionic, 64–75
  phase, 49–59
Equilibrium constant, 61
  law, 61
Ester formation, 61, 168, 176–177
Ethanal, 170–174
Ethanamide, 179, 180
Ethane, 43, 152
Ethanedioate ion, 80, 136
Ethane-1,2-diol, 157, 158
Ethanoic acid, 26, 175–178
Ethanoic anhydride, 181–182
Ethanol, 57, 165, 168, 169
Ethanonitrile, 180, 182
Ethanoyl chloride, 178, 179, 181
Ethene, 21, 22, 119, 155–158
Ethers, 169
Ethoxybenzene, 169
Ethoxyethane, 52, 55, 115, 143, 168, 169
Ethylamine, 183
Ethyne, 22, 158
Eutectic, 58–59
Exothermic, 7, 42, 83

Face-centred cube, 37
Fajans's rules, 13
Fehling's, 174, 171, 189
First order process, 19, 82, 83
Fission, 19
Flame test, 145
Fluorescence, 17
Fluorine and compounds, 99, 120–125, 126
Fractionating column, 56, 144
Fractional distillation, 56
Free energy change, 137
Free radical, 85, 192
Freezing point depression, 52–53
Friedel–Crafts, 162, 170
Fructose, 189
Fusion, 19

ΔG, 137–138
Gamma ray, 17, 18
Gas constant, 29, 54
Gases, 28–34
Gas–liquid chromatography, 144
Geiger–Müller counter, 17
Geometrical isomerism, 147–148
Germanium and compounds, 101–108
Glucose, 188, 189, 190
Graham's law, 32
Graphite, 39, 102
Ground state, 5, 17

ΔH, 41
Haber process, 63, 114, 115
Half cell, 76, 77

Half-life, 17, 83
Half reaction, 77–80
Haloform reaction, 168–169
Halogenation, 161–162, 168, 177
Halogen carrier, 161
Halogenoalkanes, 163–165
Halogenobenzenes, 163, 164
Halogens and compounds, 120–125
Heats of reaction (*see* Enthalpy)
Heisenberg uncertainty principle, 16
Helium, 125–126
Hemiacetal, 173, 189
Henry's law, 50
Hess's law, 43–44
Heterogeneous catalyst, 84
Heterogeneous equilibria, 62
Heterolytic fission, 150
1,2,3,4,5,6-Hexachlorocyclohexane, 159
Hexagonal close packing, 36
Hexane-1,6-diamine, 193
Hexanedioic acid, 193
Hexoses, 188
Hofmann degradation, 181
Homogeneous catalyst, 84
Homologous series, 149, 152
Homolytic fission, 150
Hund's rule, 3
Hybridisation, 10, 11, 21, 22
Hydration, 26
  energy, 46, 96
Hydrazine, 174
Hydrazones, 174
Hydrides, 25, 90, 93–94
Hydrogen and compounds, 92–94
  bonding, 25, 26, 27, 39, 150
  cyanide, 172–173
  peroxide, 115, 140
  sulphide, 25, 117, 118
Hydrogenation, 155
Hydrolysis, 72–73, 85–86, 117, 164, 178, 180, 181, 182
Hydroxyl group, 165–169
3-Hydroxybutanal, 172
Hydroxylamine, 173
Hydroxynitriles, 172
(4-Hydroxyphenyl)azobenzene, 186
2-Hydroxypropanonitrile, 173
2-Hydroxypropanoic acid, 173

Ice, 25
Ideal gas, 29
Ideal gas equation, 29
Ideal solution, 51
Indicators (acid-base), 73–75
Inductive effect, 12, 151, 163, 164
Inert gases, 125–126
  pair effect, 102
Infrared spectroscopy, 146
Inhibitor, 84
Intermediate compound theory, 84
Intermolecular force, 24
Interstitial compound, 128
Intramolecular force, 25
Iodine and compounds, 120–125
Iodoform (*see* Triodomethane)
Ion, 3, 7
Ionic bond, 8, 12
Ionic size, 6, 8, 87
Ionisation energy, 1, 3, 4, 7
Iron, 127, 136
  compounds, 136–139
Isobars, 14
Isoelectronic, 109
Isomerism, 146–149
Isopropyl alcohol (*see* Propan-2-ol)
Isotactic polymers, 192
Isothermal process, 47
Isotonic solutions, 54
Isotope, 14

Joule, 29

Kekulé, 22, 23, 159
Kelvin, 28
Kinetic theory of gases, 31–32
Kinetics, 81–86
Krypton, 125–126

Lassaigne test, 145
Latent heat of fusion, 35
  vaporization, 34
Lattice, 34–35
  energy, 44, 45, 46
  vibrations, 34
Le Chatelier's principle, 62–63
Lead and compounds, 101–108
Lewis acid, 65, 98, 162
  base, 65, 104
Ligand, 129
Light, 85, 153, 159
Lithium and compounds, 90, 91, 94–98, 172, 177, 180
Lyman series, 5

Macromolecule, 39, 191–194
Magnesium and compounds, 94–98
Maltose, 189, 190
Manganese, 127, 134
  compounds, 134–136
Markownikoff's rule, 156
Mass action law, 60
  defect, 18
  number, 1, 14
  spectrometer, 15, 16, 146
m/e ratio, 14, 15
Mechanisms, 85–86, 151
Melting, 35
Melting point, 45, 89
Mesomerism, 23
Meta-isomers, 160
Metaldehyde, 172
Metal extraction, 80

Metallic bonding, 36
Metalloid, 101
Methanal, 170
Methane, 10, 11, 20, 85, 104, 152–153
Methanoic acid, 175
Methanol, 166, 167
Methoxyethane, 169
Methoxymethane, 169
Methylamine, 183, 184
Methylbenzene, 160, 162, 163
Methyl-2-nitrobenzene, 160
Methyl-4-nitrobenzene, 160
Methyl orange, 73, 74
N-Methylphenylamine, 183
2-Methylpropan-2-ol, 165
2-Methylpropanoic acid, 175
Microwaves, 17
Millikan's oil drop experiment, 14–15
Molality, 51
Molecular depression constant, 52
  elevation constant, 51
  formula, 145, 146
  velocities, 33–34
Molecular weight (*see* Relative molecular mass)
Molecularity, 82
Mole fraction, 50
Monomer, 191

Naphthalen-2-ol, 186
Natural gas, 153
Neon, 125
Neutralization, 70
Neutron, 1, 15
Nickel, 127, 139
  compounds, 139, 140
Nitrate ion, structure of, 23
Nitration,
  of alkanes, 154
  of aromatic hydrocarbons, 160
Nitric acid, 114, 115
Nitriles, 182
Nitrogen, 108, 109
  compounds, 108–115
Nitrosyl cation, 110
Nitryl cation, 160
o- and *p*-Nitrotoluene (*see* Methyl-2-nitrobenzene and methyl-4-nitrobenzene)

Octet rule, 3
Optical isomerism, 148–149
Orbital, 2, 16
Order of reaction, 82
Ortho-isomers, 159
Osmosis, 53
  laws of, 54
Osmotic pressure, 54
  determination of, 54
  determination of relative molecular mass by, 54

Oxalates (*see* Ethanedioates)
Oxidation, 79
Oxidising agents, 80
Oxonium ion, 116
Oxygen, 115, 116
  compounds, 115–120
Ozone (*see* Trioxygen)
Ozonides, 157
Ozonolysis, 157

Palladium, 155
Para-isomers, 159
Paraldehyde, 172
Paramagnetic, 110
Partial covalent character, 13
  ionic character, 12, 13
  pressure, 29
Partition coefficient, 55
  law, 55
Paschen series, 5
Pauli exclusion principle, 2
Pentane, 149, 152
Pentan-2-one, 170
Peptide link, 188
Period, trends in, 6–8
Periodic Table, 1, 5–8
Periodicity, 87–91
Peroxotrifluoroethanoic acid, 157
Petroleum, crude, 153
pH, 66
Phase equilibria, 49–59
Phenetole (*see* Ethoxybenzene)
Phenol, 166, 168, 169, 179, 182, 185, 186
Phenolphthalein, 73, 74
Phenoxide ion, stabilisation of, 166
Phenyl benzenecarboxylate, 177
Phenylamine, 183
Phenylammonium chloride, 184
N-Phenylbenzenecarboxamide, 179
Phenylethanone, 170
Phenylethene, 191
Phenylhydrazine, 174
Phosphorus, 108, 109
  compounds, 108, 111, 112, 113, 114
Photochemical reactions, 85
Photon, 4
Physical properties, influence of functional group, 149, 150
Pig iron, 137
pK_a, 69
pK_b, 69
Planck's constant, 5
Plane polarised light, 148
Platinum, 77, 115, 119, 155
Polarisation of covalent bond, 12
  of ions, 13
Polyamides, 192, 193, 194
Poly(chloroethene), 191
Polyesters, 192, 193, 194

Poly(ethene), 158, 191
Polyhalide ions, 125
Polymerisation, 191
Polymers, 191–194
Polypeptides, 188
Poly(phenylethene), 191
Poly(propenonitrile), 191
Polysaccharides, 188, 190
Positron, 18, 19
Potassium, 94
  extraction, 80
  compounds, 94–97
Propanal, 170
Propane, 152
Propanoic acid, 176
Propanonitrile, 182
Propan-2-ol, 165
Propanone, 24, 170, 171
Propene, 155
Proteins, 25, 188
Proton, 1, 15
Purification of organic compounds, 142–144

Quartz, 106
Quaternary ammonium salts, 182, 183

Racemic mixture, 148
Radioactivity, 16–19
Radon, 125
Raney nickel, 155
Raoult's law, 50
Rast's method, 53
Rate constant, 81
  determining step, 82
  of reaction, 81–85
Reaction,
  endothermic, 42
  exothermic, 42
  mechanism, 85–86, 151
  molecularity of, 82
  order of, 82
Real gases, 33
Redox reactions, 79–80
Reducing agents, 80, 93, 94
Reduction, 79
Relative atomic mass, 14
Relative molecular mass, determination of, 15, 50–54
Resolution, 148
Resonance, 22–23
Resonance hybrid, 23
R_f value, 144
Root mean square velocity, 31
Rusting, 79, 138
Rutherford, 15
Rydberg constant, 5

Salt bridge, 77
Salting out, 65
Saturated compounds, 153
Scandium, 127, 128, 129
Schiff's base, 174
Schrödinger wave equation, 16
Semicarbazide, 174
Semipermeable membrane, 53
Shapes of molecules and ions, 20–23
Silicon and compounds, 101–108
Silver and compounds, 120, 124, 125, 130, 158
Smelting, 136
Soda lime, 153
Sodium, 94
  extraction, 80
  compounds, 94–97
Solubility, factors affecting, 13, 26, 46, 47
Solubility product, 64
Solutions of gases in liquids, 50
  liquids in liquids, 55–57
  solids in liquids, 50–55
Solvation, 26
Solvent extraction, 55, 142
Spontaneous reaction, 47–48
Stability constant, 130–131
Standard electrode potential, 76, 78
Starch, 121, 190
States of matter,
  gases, 28
  liquids, 34
  solids, 34
Steam distillation, 57–58, 144
Steel, 137
Stereoisomerism, 147–149
Steric hindrance, 171
Structural formula, 145
  determination of, 146
Styrene (*see* Phenylethene)
Sublimation, 35
Sucrose, 189–190
Sugars, 188–190
Sulphonation,
  of alkanes, 153
  of aromatic hydrocarbons, 161
Sulphur, 115, 116
  compounds, 115–120
Sulphur dichloride oxide, 178
Sulphuric acid, 119–120
  as a dehydrating agent, 119, 167
  as a catalyst, 168, 176, 177
Syndiotactic polymers, 192

Terephthalic acid (*see* Benzene-1,4-dicarboxylic acid)
Terylene, 193
Tetrachloromethane, 85, 107, 153
Thermite process, 100
Thermodynamics,
  First law, 41
  Second law, 47
Thionyl chloride (*see* Sulphur dichloride oxide)

Thomson, 14
Tin and compounds, 101–108
Titanium, 127, 131
  extraction, 80
  compounds, 131–132
Titrations, acid-base, 73–75
Tollens's reagent, 171
Toluene (*see* Methylbenzene)
Trans-isomers, 147
Transition elements, 127–141
Transition state, 83
Translational energy, 33
2,4,6-Tribromophenol, 169
Trichloroethanoic acid, 176, 177
Trichloromethane, 56, 85, 153
(Trichloromethyl)benzene, 162
Trimethylamine, 183
Triodomethane, 168, 171
Trioxygen, structure of, 116
Tritium, 14

Ultraviolet light, 17, 85, 159
Unimolecular reaction, 86
Unit cell, 37
Unsaturated compounds, 154
Uranium, 18, 19

Van der Waals forces, 24, 125
Vanadium, 127, 132
  compounds, 132–133
Van't Hoff factor, 55
Vapour density, 30
Vapour pressure, 49, 55–58
  lowering of, 50
Velocities of molecules,
  distribution in a gas, 33
Vinyl chloride (*see* Chloroethene)
Viscosity of alcohols, 150

Water, 116
  amphiprotic nature, 70
  as a solvent, 26
  hydrogen bonding in, 25
  ionic product of, 66
  shape, 20, 21
Weak acids and bases,
  strengths of, 68
Williamson's synthesis, 165

Xenon, 125
  compounds, 126
  tetrafluoride (shape), 21
X-rays, 17
X-ray diffraction, 35

Zartmann experiment, 33
Ziegler–Natta catalysts, 128, 192
Zinc, 127, 128, 129
  extraction, 80
  sulphide lattice, 38
Zwitterion, 187